Small Unmanned Aerial System Adversary Capabilities

BRADLEY WILSON, SHANE TIERNEY, BRENDAN TOLAND, RACHEL M. BURNS, COLBY PEYTON STEINER, CHRISTOPHER SCOTT ADAMS, MICHAEL NIXON, RAZA KHAN, MICHELLE D. ZIEGLER, JAN OSBURG, IKE CHANG

Published in 2020

Preface

This report describes current and projected performance capabilities and limitations of small unmanned aerial systems (sUASs), also known as drones. Specifically, it provides a component-level assessment of sUAS technologies.

The analysis captured the potential capabilities and limitations of these platforms, with the results of the analysis intended to help the program office inform strategic policy and investment decisions to counter UAS threats to prevent strategic surprise. This document is intended to be updated at least annually by the U.S. Department of Homeland Security Science and Technology Directorate's Program Executive Office for Unmanned Aerial Systems.

This research was sponsored by the U.S. Department of Homeland Security (DHS) Science and Technology Directorate and conducted within the Acquisition and Development Program of the Homeland Security Operational Analysis Center (HSOAC) federally funded research and development center (FFRDC).

About the Homeland Security Operational Analysis Center

The Homeland Security Act of 2002 (Section 305 of Public Law 107-296, as codified at 6 U.S.C. § 185) authorizes the Secretary of Homeland Security, acting through the Under Secretary for Science and Technology, to establish one or more FFRDCs to provide independent analysis of homeland security issues. The RAND Corporation operates HSOAC as an FFRDC for DHS under contract HSHQDC-16-D-00007.

The HSOAC FFRDC provides the government with independent and objective analyses and advice in core areas important to the department in support of policy development, decisionmaking, alternative approaches, and new ideas on issues of significance. The HSOAC FFRDC also works with and supports other federal, state, local, tribal, and public- and private-sector organizations that make up the homeland security enterprise. The HSOAC FFRDC's research is undertaken by mutual consent with DHS and is organized as a set of discrete tasks. This report presents the results of research and analysis conducted under 70RSAT18FR0000037, Unmanned Aerial Systems Adversary Capabilities.

The results presented in this report do not necessarily reflect official DHS opinion or policy. For more information on HSOAC, see www.rand.org/hsoac. For more information on this publication, visit www.rand.org/t/RR3023.

Contents

Figures

Tables

Summary

Purpose and Approach

The U.S. Department of Homeland Security (DHS) Science and Technology Directorate (S&T) asked the Homeland Security Operational Analysis Center to create an unmanned aerial system (UAS) adversary capabilities document to assess the capabilities of small UASs[1] (sUASs) that may be used for nefarious purposes in the next five to ten years. This assessment employed a four-step process:

1. Perform a component-level assessment of sUAS technologies.
2. Complete a tradespace analysis of sUAS performance characteristics.
3. Assess potential nefarious uses of sUASs.
4. Assess the challenges that sUAS presents to counter-UAS (C-UAS) sensors.

Information and descriptions of current and future sUAS technology and limitations were gathered from a multitude of sources. The Association for Unmanned Vehicle Systems International Unmanned Systems and Robotics Database–Air Platforms served as a primary source for current and historical UAS platforms. We also drew on *Jane's Defence: Air Platforms* resources, manufacturer websites, academic papers, and corporate white papers.

Our study also developed a framework for how nefarious actors might exploit sUAS technologies. We drew on DHS working group reports, open-source research, and literature reviews to create additional possible use cases. These use cases were informed by recent current events, including the swarming event reported by Federal Bureau of Investigation agents in May 2018 and the August 2018 drone assassination attempt against Venezuelan President Nicolas Maduro. Using these resources, we generalized the various use cases and identified a subset of notional cases that are highly likely, highly damaging, or both.

---

[1] A *small* UAS is defined by the Federal Aviation Administration (FAA) as a UAS with a maximum gross takeoff weight of 55 lbs or less (FAA, 2018).

The goal is to inform DHS on areas of investment, challenges, and capabilities for C-UAS. Countering sUASs is harder than countering larger drones for obvious reasons: Smaller drones are generally more difficult to detect, classify, and identify, particularly in regions with high levels of clutter, such as urban areas. In some circumstances, it can be difficult to distinguish a drone from a bird, and sensors, such as radar, may be restricted so as not to interfere with airport operations or other activities.

What We Found

Consequential Nefarious sUAS Uses Are Possible—and Even Likely—with Today's Technologies

We identified four high-risk notional use cases involving the nefarious use of sUASs for

- unauthorized reconnaissance or surveillance
- conveying illicit material
- conducting a kamikaze explosive (i.e., kinetic) attack
- conducting a chemical, biological, or radiological attack.

The four notional use cases developed for this study could be carried out using systems that are readily available through the commercial market. At least 25 percent (351 of 1,429) of the systems in the database could be used in one of the four use scenarios. At least 27 platforms on the market today could execute all four; however, each system is specifically marketed and sold for national defense uses and therefore may not be available to retail consumers. Our analysis led us to several general observations:

- Nearly a quarter of the systems surveyed could complete our notional surveillance and reconnaissance scenario, requiring an hour or more of flight time.
- More than 300 surveillance and reconnaissance systems available today could meet the requirement of continuous flight for an hour or longer; 50 of them are available on the retail commercial market. These sUASs are predominately fixed-wing airframes with battery electric propulsion.
- Fifty systems available today are capable of our notional conveyance scenario—primarily fixed-wing airframes with internal combustion engines. However, a handful of rotary-wing models would suffice. This assumes a range greater than 20 miles, flight time of an hour or longer, and payload greater than 10 lbs.
- Due to the high speed of our notional kamikaze attack scenario, it is more the domain of fixed-wing airframes. There are 70 options, of which only nine are considered retail commercial. This assumes that a maximum speed of greater than 75 knots is preferred to reduce the potential reaction time of C-UAS capabilities, as well as a payload weight of 2.2 lbs or more.

- Finally, only some systems are capable of carrying out an unlikely but highly consequential notional aerosolized attack with a chemical, biological, or radiological agent as described in our notional scenario, which requires a payload greater than 10 lbs, the ability to fly for at least 30 minutes, and a range of at least five miles. There are about 40 fixed-wing and 40 rotary-wing systems that can complete such a mission. If we consider the retail commercial market only, there are eight fixed-wing and five rotary-wing sUASs with this capability.

The Problem Will Get Worse

Market estimates indicate that Chinese firm Dà-Jiang Innovations (DJI) accounts for 70 percent of the sUAS retail market. Trends in DJI system performance indicate capability advancement. For example, the DJI Phantom's endurance has increased by about 1.5 minutes per year, and its maximum speed has increased 1.7 m/s over the past four years. Whether this will continue in the future is unknown. However, there is reason to believe the problem will only get worse:

- Miniaturized radar, hyperspectral, and light detection and ranging payloads have been developed to work onboard sUASs. The Walden curve shows a tenfold increase in the processing capability of analog-to-digital converters, enabling sUASs to collect more electronic signals.[2]
- Global Navigation Satellite System jamming and spoofing will continue to be attractive applications for nefarious sUAS users due to the limited power required to jam or spoof and the ability of an sUAS to get into unobstructed positions to maximize its impact.
- Much of the command-and-control market is focused on increasing the autonomy of flight software using artificial intelligence, predictive analytics, and computer vision to reduce the requirement need for a highly trained operator.
- There are many how-to guides on the internet describing how to fit an sUAS with a cellular network transceiver for remote control. Users are likely dealing with latency on today's 4G (fourth-generation) network, but advances in this area will make it more difficult to identify sUASs via communication links.
- There is a robust open-source software and hardware ecosystem for supporting autonomous control, autopilot, and other sensors and control systems needed for remote operation.
- Increased autonomous capabilities and a desire for faster and simpler control are pushing manufacturers toward specialized controllers, often with separate controls for flight and payload. Autonomous technologies are reducing the complexity and need for operator training.

[2] The Walden curve describes the performance of systems, such as UAS sensors, that convert analog signals into digital signals.

- Battery technology is notorious for overpromising, but we estimate that flow batteries could improve the range of the DJI Phantom 4, for example, by 20 percent, and solid-state batteries by 100 percent.

Common characteristics of the sUAS fleet being manufactured today appear to include the following:

- reliance on lithium-polymer batteries
- prevalence of carbon fiber composite airframes
- the use spread spectrum frequencies for command and control of the platforms
- a payload weight of about 25–40 percent of the maximum gross takeoff weight for most fixed- and rotary-wing UASs.
- advancing technology in hybrid-wing hybrid-propulsion platforms, providing these systems with the longest range (419 miles) and endurance (520 minutes) of the sUAS platforms considered in this study
- rotary-wing internal combustion platforms, which have the highest mean payload weight at 13.6 lbs.

Considering the increasing payload on performance, we found that fixed-wing sUASs—especially those with internal combustion engines—were the leaders in terms of range, endurance, and speed.

Two findings indicate that the potential problem of nefarious sUAS use may not be as bad as it seems:

- Top-selling sUASs have lower-than-average endurance (mean of 21 minutes versus 33 minutes for non–top sellers)—lower than the average maximum range (mean of 1.8 miles versus 4.7 miles) and lower than average payload weights (4.6 lbs versus 7.4 lbs). We found no differences in average speed or average maximum altitude.
- sUASs with a stated intent of serving as prototypes were not statistically better or worse in any of the five performance categories studied than sUASs with any other stated intent.

Performance characteristics have trade-offs. Increasing one capability comes at a cost to another: Increasing payload reduces endurance, and reducing size limits sensor selection, all other things being constant. In examining these trade-offs, we found that fixed-wing sUASs—especially those with internal combustion engines—were the leaders in terms of range, endurance, and speed. However, rotorcraft can carry similar payloads, and their reduced detection signatures and ability to hover and perform agile maneuvers make them an attractive option for nefarious users.

In analyzing the trade-offs among performance characteristics, we found that fixed-wing sUAS have range, speed, and endurance advantages over rotary-wing sUASs. However, hybrid designs that combine both methods of lift generation may offer a combination of improved range and endurance and improved maneuverability and station-keeping. Internal combustion engines offer improved power, leading to superior performance (e.g., speed, endurance, payload) but also increased complexity. In particular, they tend to have greater mechanical complexity, a larger acoustic signature, and more-complex ways of interfacing with their onboard electronics. Furthermore, internal combustion engines have diminishing returns at very small sizes. These characteristics explain why many manufacturers use battery electric power instead of internal combustion engines. This also implies that electric-powered systems will very likely fall short of combustion engine systems in terms of range, endurance, and payload.

Implications for Counter-sUAS Efforts

The ability to detect an sUAS depends on the sensors doing the detecting, and each of these sensors comes with trade-offs. SUASs present unique challenges to C-UAS systems due to their size and operational use (among other things), and detecting, classifying, identifying, and tracking sUASs often requires numerous sensors working in concert:

- The smaller the sUAS, the more problematic it is for C-UAS, because it is more likely to generate false alarms. This can play havoc on the human operators monitoring the sensors and systems, desensitizing them to real threats.
- The radar cross-section (RCS) of rotary-wing sUASs can vary by several orders of magnitude, depending on the aspect angle of the sensor. This necessitates varied radar configurations, such as the use of multiple radars or a single radar with wider bandwidth.

DHS should be aware that not all military grade C-UAS systems are suitable for countering sUASs:

- Most military-grade target-tracking radars operate in the X-Band (8–12 GHz), where antenna gain is high enough for good angular resolution, and, at the same time, the attenuation is weak enough to allow target tracking at ranges of hundreds of kilometers. But, against the sUAS threat, candidate radars do not have to detect and track at such long ranges, making higher-frequency bands (e.g., Ku 12–18 GHz, Ka 26.5–40 GHz) more attractive.
- Trade-offs in the performance of radar and imaging sensors will make C-UAS procurement of more difficult, particularly if one-size-fits-all solutions are desired. For example, although higher-frequency bands are more suitable for detecting smaller UASs, they are more sensitive to poor weather conditions.

- Due to such natural trade-offs in imaging sensors, those operating at the higher resolution necessary to detect sUASs at range are likely limited to a narrow field of view. This necessitates a systems-of-systems approach to search for and detect targets of interest using wide-area/lower-resolution sensors and a cue to higher-resolution sensors to track and identify them, requiring timely coordination between the various sensors.

Efforts by legitimate sUAS users to reduce their platforms' size and potential for interference to avoid being a nuisance may be a boon for nefarious users and necessitate investment in capable and robust detection technologies and approaches:

- The commercial market is moving toward reduced acoustic signatures, which will continue to pose problems for acoustic detection that is already limited to close ranges.
- Although reduced RCS and command-and-control signal obfuscation may not be explicitly sought by lawful operators (and therefore market forces), they may appear as byproducts of other developments (e.g., airframe material or aircraft design and miniaturization unintentionally obfuscating RCS or cell signals).

This report and its underlying data are intended to be foundational and updateable by S&T as technological capabilities evolve, new insights on nefarious use cases are collected, and the impact on C-UAS changes. For these reasons, the performance analysis in Chapter Three was designed to be repeatable.

Acknowledgments

We thank the project's sponsor, Jeffrey Randorf of DHS S&T, as well as his colleague Ralph Gibson. We are also grateful to former RAND colleague Colin Ludwig for his contributions. Thank you to our reviewers, Paul Dreyer and John Parmentola, and the project's administrative assistant, Laura Coley. Lauren Skrabala provided numerous reviews and edits, improving the overall quality of the document. Dan Spagiare reviewed an early draft and helped improve the overall format.

Abbreviations

3D	three-dimensional
A/D	analog to digital
ACD	adversary capabilities document
ADS-B	automatic dependent surveillance-broadcast
AI	artificial intelligence
AQIM	Al Qaeda in the Islamic Maghreb
AUVSI	Association for Unmanned Vehicle Systems International
C3	command, control, communication
C-UAS	counter–unmanned aerial system
CBP	U.S. Customs and Border Protection
CBR	chemical, biological, and radiological
CONOPS	concept of operations
CPU	central processing unit
CW	continuous wave
D/A	digital to analog
DARPA	Defense Advanced Research Projects Agency
DHS	U.S. Department of Homeland Security
DJI	Dà-Jiang Innovations
DRFM	digital radio frequency memory
ELINT	electronic intelligence

EM	electromagnetic
EO	electro-optical
EO/IR	electro-optical/infrared
ERP	effective radiated power
ESM	electronic support measures
EW	electronic warfare
FAA	Federal Aviation Administration
FAR	false alarm rate
FFRDC	federally funded research and development center
FOV	field of view
FOR	field of regard
FPGA	field-programmable gate array
FPS	Federal Protective Service
GFLOPS	giga–floating point operations per second
GMTI	ground moving-target indicator
GNSS	global navigation satellite system
GPS	Global Positioning System
GPU	graphical processing unit
HSOAC	Homeland Security Operational Analysis Center
IQR	interquartile range
IR	infrared
IRST	infrared search and track
ISR	intelligence, surveillance, and reconnaissance
J/S	jammer-to-signal ratio
Li-air	lithium-air
LIDAR	light detection and ranging

Li-ion	lithium-ion
Li-metal	lithium-metal
LiPo	lithium-polymer
Li-S	lithium-sulfur
LWIR	long-wave infrared
MDV	minimum detectable velocity
MGTOW	maximum gross takeoff weight
MOMU	multiple operators with multiple unmanned aerial systems
MWIR	midwave infrared
NASA	National Aeronautics and Space Administration
NSSE	national security special event
OPS	Office of Operations Coordination
PRF	pulse repetition frequency
RC	radio control
RCS	radar cross-section
RF	radio frequency
S&T	Science and Technology Directorate
SAR	synthetic aperture radar
SATCOM	satellite communication
SDR	software-defined radio
SEAR	Special Event Assessment Rating
SESAR	Single European Sky Air Traffic Management Research
SMRF	scalable multifunction radio frequency
SNR	signal-to-noise ratio
SSA	Social Security Administration

STOL	short takeoff and landing
sUAS	small unmanned aerial system
SWaP	size, weight, and power
SWIR	short-wave infrared
TDOA	time difference of arrival
UAS	unmanned aerial system
USBP	U.S. Border Patrol
USCG	U.S. Coast Guard
USSS	U.S. Secret Service
UTM	unmanned aerial system traffic management
VHF	very high frequency
VOR	very high frequency omnidirectional range
VTOL	vertical takeoff and landing

CHAPTER ONE

Introduction

Background and Purpose

The U.S. Department of Homeland Security (DHS) Science and Technology Directorate (S&T) asked the Homeland Security Operational Analysis Center (HSOAC) to create an unmanned aerial systems (UAS) adversary capabilities document (ACD). An ACD is an estimate of how adversary capabilities could evolve over a given period (Montroll, 2003). Specifically, we assessed the capabilities of small UASs (sUASs) that may be used for nefarious purposes in the next five to ten years by surveying UAS technology and applications, forecasting their development, and identifying trends in "possibly novel and and nefarious application of these technologies within the homeland security context."[1]

The goal of this capability assessment is to inform DHS strategy and research and development initiatives as the department prepares to meet future counter-UAS (C-UAS) requirements.

The commercial sUAS market continues to grow. Platform costs have come down, performance capabilities continue to increase, barriers to entry are lowering, do-it-yourself construction is more practical, and innovative ideas are emerging as sUASs are used to satisfy a myriad of needs, including imaging, disaster response, search and rescue, security, cable services, agriculture, and site inspection.

Our study involved a four-step approach:

1. Perform a component-level assessment of sUAS technologies.
2. Complete a tradespace analysis of sUAS performance characteristics.
3. Assess the potential nefarious uses of sUASs.
4. Assess the challenges that sUASs present for C-UAS sensors

Each step listed above corresponds to a chapter in this report.

[1] A *small* UASs is defined by the Federal Aviation Administration (FAA) as a UAS with a maximum gross takeoff weight (MGTOW) of 55 lbs or less (FAA, 2018).

Performance Characteristics and Component Areas

There are multiple ways to assess the performance of sUASs. Common ways that manufacturers report performance include listing range, endurance, payload, and speed. However, this reporting is not standard across manufacturers and requires normalization. For example, it can be unclear whether a reported range includes a return trip.

The component areas studied are as follows:

- airframes
- payloads and sensors
- command, control, communication (C3)
- power supplies and propulsion
- software security
- detectability.

Study Development

This report and its underlying data are intended to be foundational and updateable by S&T. Deeper understanding of technology capabilities, new insights on nefarious use cases, and the impact on C-UAS can continue to be built upon. The performance analysis in Chapter Three relied on easily accessible data and was designed to be repeatable.

This process was conceived of in three phases:

- Phase 1
  - Data collection: Build a consolidated data set of reported performance of commercial sUASs.
  - Historical data collection: Expand the consolidated data set with performance data over time to identify performance trends.
  - Analysis: Identify nefarious uses of sUASs.
  - Analysis: Describe reported performance characteristics (including trend analysis).
- Phase 2
  - Data collection: Build a data set of design characteristics (i.e., the components of sUASs and their performance).
  - Analysis: Incorporate existing and develop new models of performance for sUASs.
  - Analysis: Use models to predict performance and compare reported performance.
  - Analysis: Expand component-level analysis by creating hypothetical combinations of components to explore the capability of hypothetical systems.

- Phase 3
  - Data collection: Identify nefarious uses of sUASs and capture performance requirements (e.g., payload demands to carry out certain types of nefarious missions).
  - Analysis: Connect the phase 2 component-level analysis to the nefarious use requirements and determine the types and availability of systems and components that can execute nefarious acts.

This report describes the phase 1 data collection and analysis, along with some phase 2 component-level modeling and analysis and some phase 3 identification of performance capabilities to support DHS efforts to counter nefarious uses. We recommend that future work update the phase 1 data, expand the model and component analysis in phase 2, look more holistically across the nefarious use cases, and assess in greater detail the performance demands of the use cases in phase 3.

Data-Gathering Methodology

We gathered a significant amount of information on current and future sUAS technology and its limitations from a multitude of sources. The Association for Unmanned Vehicle Systems International's (AUVSI's) Unmanned Systems and Robotics Database–Air Platforms served as a source for data on current and historical UAS platforms. We also drew from *Jane's Defence: Air Platforms*, manufacturer websites, and academic papers. Table 1.1 lists the manufacturers whose websites we consulted to build our data set, along with the systems that were of particular interest for this study. Specific sources cited in this report can be found in the References section.

We consulted a variety of models, reviewed corporate white papers, and developed a set of representative equations to assess near-term (five to ten years) limitations. We consolidated our findings into a database that identifies key physical and performance metrics of each system and enables a detailed analysis of current and future trends in the sUAS industry. The information we collected on current and historical threats allowed us to identify some general trends. These trends, while not engineering laws, facilitated educated predictions regarding what threats can be expected from future sUAS equipment.

Scoping the sUAS Market

The sUAS market is broad. In total, we identified more than 600 manufacturers of over 1,700 discrete systems across 28 different stated intents (that is, the stated purpose of a given system). Stated intents are discussed further in Chapter Three (see Figures 3.11 and 3.12) and include imaging, disaster response, precision agriculture, mining, prototypes, attack, and weapon delivery.

Table 1.1
Platforms and Manufacturers in the Data Set

Company	Platform(s)
Airborne Drones	The Vanguard
Alpha Unmanned	Alpha 800
Autel Robotics	EVO
Dà-Jiang Innovations (DJI)	Mavic 2 Enterprise
HexaMedia	ZionPro520 (Quad)
Impossible Aerospace	IMPOSSIBLE US-1
Robots in Search	SteadiDrone EI8HT RTF
UCONSYSTEM	REMO M-001, REMO M-002, 5G/LTE Drone, REMOCOPTER-001, REMOCOPTER-004, Multi-Purpose Commercial Drone, Parcel Delivery Drone, REMOFARM-20, REMOEYE-002B, REMOEYE-006A, and T-ROTOR
Wingtra	WingtraOne
Ziyan UAV	Parus and Ranger

This study focused on commercial sUASs under 55 lbs, excluding prototypes but including systems designed for defense markets.[2] Table 1.2 shows the numbers of platforms and manufacturers in our data set meeting each of these criteria.

Platforms may be, for example, fixed-wing, rotary-wing, hybrid, or lighter than air. Of the 625 sUAS manufacturers, 46 percent had one platform in the data set, representing 18 percent of the total number of platforms. Table 1.3 shows the 625 sUAS manufacturers with ten or more models in the data set.

Table 1.2
Number of Platforms and Manufacturers in the Data Set

Category	Total	< 65 lbs MGTOW	Not Classified as a Prototype[a]	No Stated Defense Intent[b]
Platforms	1,733	1,579	1,429	491
Manufacturers	668	625	564	256

[a] We ultimately dropped prototype models from our analysis, as discussed in Chapter Three.

[b] *Defense intent* is defined as a stated intent of electronic warfare; intelligence, surveillance, and reconnaissance (ISR); targeting; target acquisition; or attack.

[2] Prototype performance was not significantly better or worse than that of non-prototypes; we left prototypes out of the analysis in Chapter Three because they were not commercially available. Chapter Three uses 65 lbs as the MGTOW threshold, as reported data seemed to vary across the industry.

Table 1.3
Manufacturers of sUAS in the Consolidated Study Data Set

Manufacturer (Country)	Number of Platforms	Number of Platforms with No Defense Intent	Number of Platforms with Defense Intent (% of manufacturer's platforms)
AeroVironment (France)	24	5	19 (79)
DJI (China)	24	18	6 (25)
U.S. Naval Research Laboratory (United States)	16	8	8 (50)
UCONSYSTEM (South Korea)	15	4	11 (73)
Zala Aero Group (Russia)	13	1	12 (92)
Israel Aerospace Industries (Israel)	11	0	11 (100)
Lockheed Martin Corporation (United States)	11	1	10 (80)
Draganfly Innovations (United States)	10	2	8 (80)
ZEROTECH (China)	10	6	4 (40)

Of the 625 sUAS manufacturers, approximately 300 produced only one model, as shown in Figure 1.1.

We did not specifically analyze the models produced by single-platform manufacturers. However, we identified a significant number of prototypes in this category, and

Figure 1.1
Frequency of sUAS Manufacturers in the Data Set

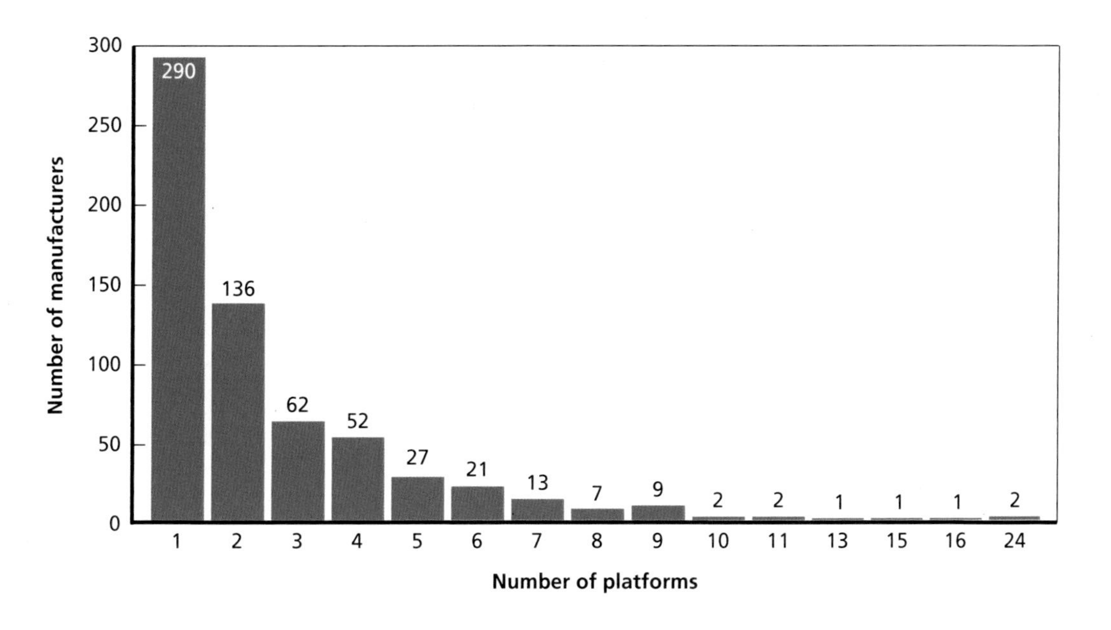

these were much more likely to be categorized as prototypes (17.6 percent) than those manufactured by companies in our database that produced more than one platform (7.7 percent).

The market (as determined by sales) is dominated both globally and within the United States by the Chinese manufacturer DJI's rotorcraft. A third-quarter 2016 assessment by Drone Industry Insights (Drone II) identified DJI as the number 1–ranked company as measured by Google searches, news items, and employees, followed by French manufacturer Parrot.

It is important to note that the scores in Figure 1.2 are not a reflection of market share and that the rankings have changed since the time of this research. *TechNode* reported in January 2018 that "the Shenzhen-based drone maker is currently the world's top seller of consumer drones, with a global market share of 70%" (Borak, 2018). Chapter Three discusses the market and capabilities of the 1,429 sUAS non-prototype models in greater detail.

Figure 1.2
Drone II Company Rankings, Third Quarter 2016

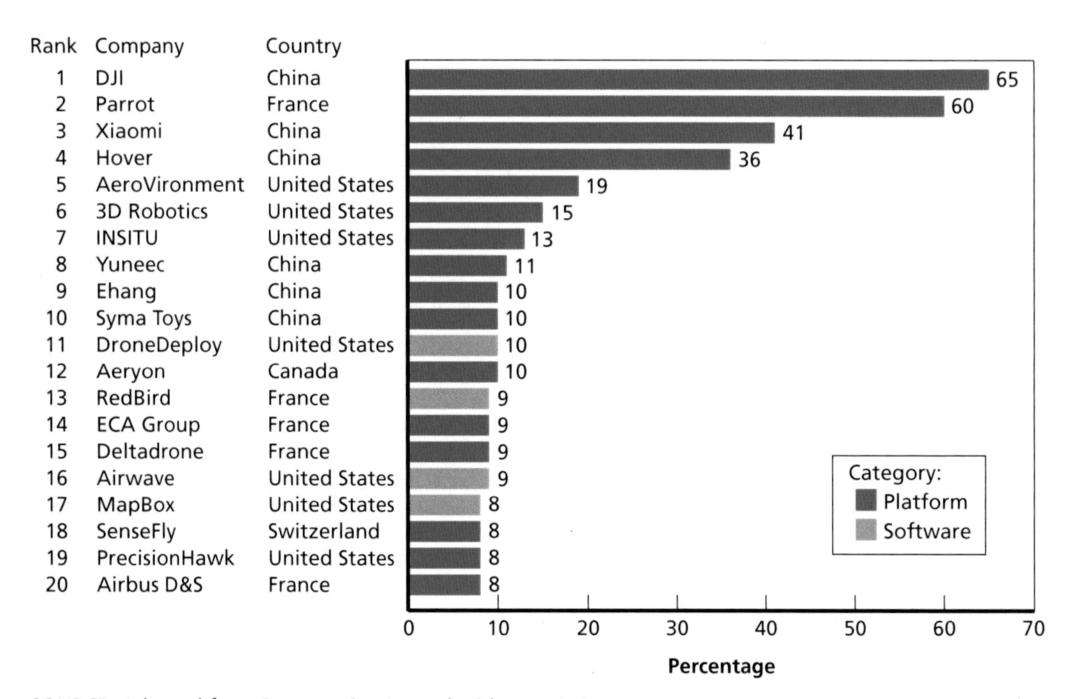

SOURCE: Adapted from Drone II, 2016. Used with permission.
NOTE: Data compiled based on number of Google searches for company names in conjunction with the keywords "drone" and "UAV," news items citing company and "drone" and "UAV," and a company employee count on LinkedIn accompanied by "drone" or "UAV." The highest-scoring company according to each metric is given a rating of 100%, with the others ranked in relation to it. The percentages represent the average of the scores.

Commercial Availability

Commercial sUASs may be categorized by the ease with which a user can obtain them. These categories can be used to measure risk of exposure incurred by a nefarious actor who obtains an sUAS:

- Some platforms are available at brick-and-mortar big-box retail stores (e.g., Best Buy, Walmart). These devices tend to be simple toys or very common hobbyist models and can be purchased with cash on the spot.
- A larger proportion of these toy and low-end hobbyist platforms can be purchased from online retailers, but a mailing address is required.
- More-specialized platforms are available at high-end hobby stores, either brick-and-mortar or online. Again, online stores require a mailing address, and high-end hobby stores often require personal interaction with employees, increasing risk to nefarious operators.
- General industrial suppliers may offer highly capable sUASs that are specialized for complex industrial demands, but access to these suppliers is often limited high costs or the need for extensive interpersonal contact with suppliers and detailed recordkeeping requirements.
- Suppliers that serve specific industries (for example, the agricultural UAS market) can also provide sUASs with better or more-desired capabilities than general retailers, but access barriers are similar to those for general industrial suppliers. In addition, these communities tend to be small and highly specialized, so unusual purchasing inquiries or activity is likely to arouse suspicion.

Do-It-Yourself Capabilities

The possibilities for amateur sUAS development are rapidly expanding. Previously, at minimum, hobbyists needed the ability to purchase and correctly assemble a premade parts using a kit, but this still required some rudimentary knowledge of model aircraft and aeronautical engineering (plus some level of computer programming and electrical engineering expertise if automation was involved). With the explosion of online communities and learning resources, component and subsystem availability through online retailers, and the dramatic drop in cost and knowledge barriers to entry for additive manufacturing—accompanied by increased material choice—have made do-it-yourself a viable means for assembling and designing sUASs (Smith, 2018).

Systems obtained in this manner generally lack the paper trail of commercially purchased platforms and allow rapid experimentation and growth of ideas. In the hands of nefarious users, this easy sharing of information and materials will allow the rapid development of systems optimized for nefarious uses that might not otherwise arise from the commercial sector (e.g., radar reflective shapes and coatings, platforms optimized for kamikaze tasks, optical or acoustic camouflage).

DJI-Manufactured Systems

Given the dominance of DJI models in the market space, we gave a closer look to their evolution and current capabilities to collect insights for studying performance trends. The DJI Phantom is the most ubiquitous sUAS model in circulation. However, the DJI Mavic and Spark are also growing in popularity in the hobbyist category of the sUAS market space. DJI's professional and enterprise models, such as Inspire, Matrice, and Agras MG, are becoming more common in the industrial category.

As the oldest DJI model, the Phantom has a significant amount of historical data from which to characterize the evolution of sUAS performance parameters. Since its introduction in 2013, the Phantom has seen little change in its size and weight, which have stabilized at a diagonal length of 350 mm and a mass of 1,375 g, as shown in Figure 1.3. However, in examining other DJI systems that have become available since 2016, we see an overall trend toward smaller systems. While this trend may be a simple coincidence, it may also imply a market shift toward smaller, lighter, and more-difficult-to-detect systems.

The Phantom's performance has improved significantly over the past six years while its relative size and weight have held steady. Figure 1.4 shows an increase in the maximum flight time of approximately 1.5 minutes per year, with a current maximum

Figure 1.3
Weight of Various DJI Phantom Models, 2013–2018

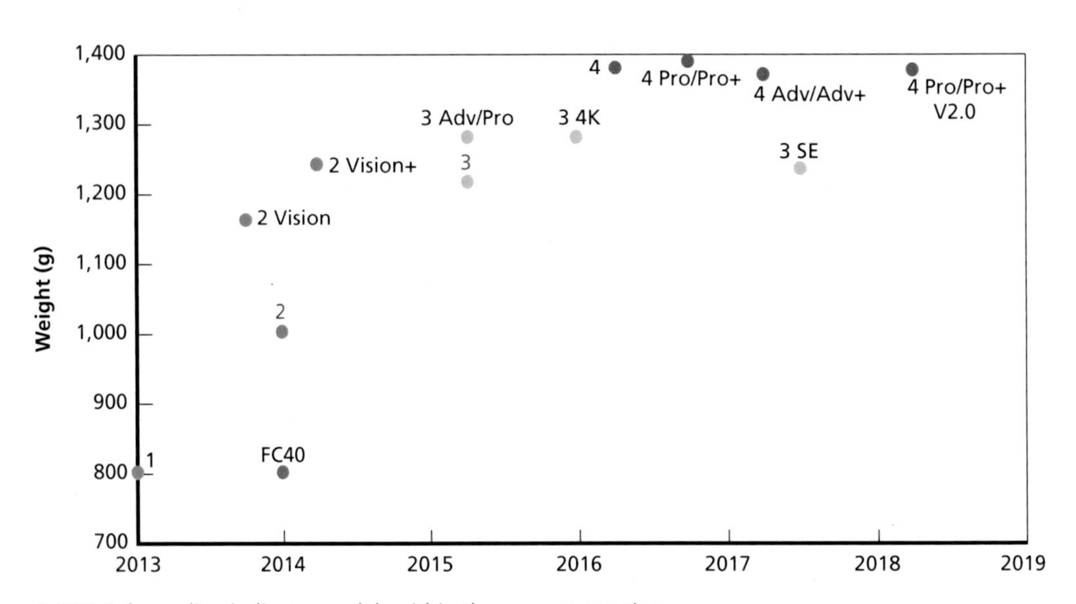

NOTE: Color-coding indicates models within the same generation.

Figure 1.4
Maximum Flight Time of Various DJI Phantom Models, 2013–2018

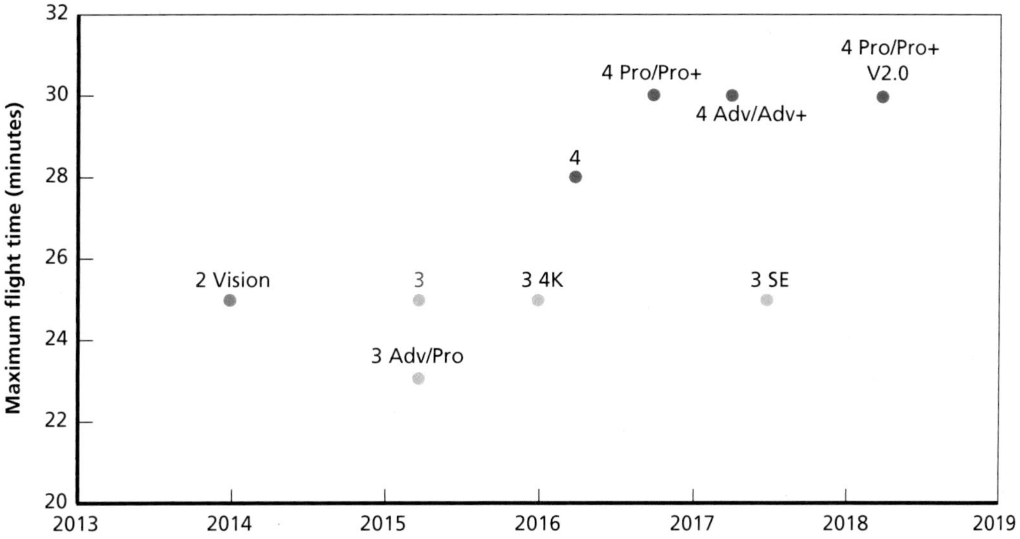

NOTE: Color-coding indicates models within the same generation.

flight time of 30 minutes for the Phantom 4 models.[3] A similar trend is clear for many of the other small DJI models available to consumers, with the exception that the Spark is capable of half the flight time of the other models, but it is also significantly smaller (20 percent of the Phantom 4's weight).

There was an increase in battery capacity from the version 1 to version 2 models, a decrease for version 3, and an increase for version 4. Anecdotal evidence has suggested that the version 3 models were able to maintain flight time with improved motor efficiency.

Similarly, the Phantom has experienced an increase in maximum speed of approximately 1.7 m/s per year, with a current maximum speed of 20 m/s for the Phantom 4 models, as shown in Figure 1.5. The clear capability increase is apparent, with the Mavic 2 systems able to maintain the Phantom 4's maximum speed at 65 percent of the Phantom 4's overall weight.

The maximum communication distance of the Phantom has also increased at a rate of approximately 1,400 m per year, with a current maximum of 7,000 m for the Phantom 4 models, as shown in Figure 1.6.

Over the course of two years, DJI was able to develop a 35-percent smaller system with specifications that are equivalent to—or even exceed—the Phantom 4, which required approximately 3.5 years of development. Our evaluation relied on a limited

---

[3] Data were not readily available on the flight times of the version 1 and FC40 models.

Figure 1.5
Maximum Speed of Various DJI Phantom Models, 2013–2018

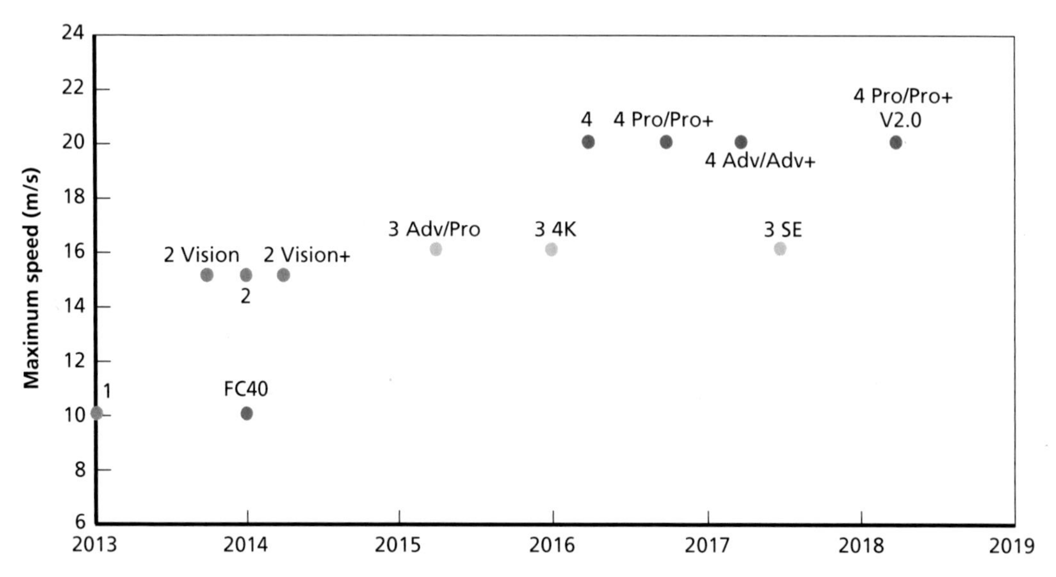

NOTE: Color-coding indicates models within the same generation.

Figure 1.6
Maximum Communication Distance of Various DJI Phantom Models, 2013–2018

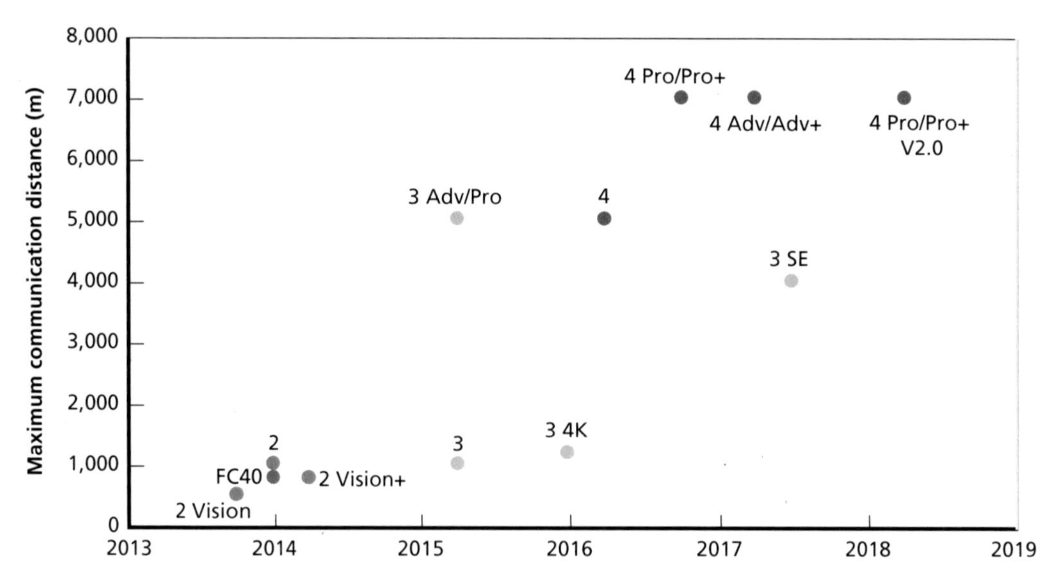

NOTE: It is unclear what is causing the increase in communication range. Color-coding indicates models within the same generation.

sample set, but the results imply that sUAS capabilities can be expected to increase more rapidly going forward than in previous years. As the new Spark system continues to mature, we can expect to see significant increases in its capabilities as well.

Compared with the Phantom, less historical data are available on the performance parameters of DJI's industrial sUAS models. However, a few important insights can be gleaned. First, the Matrice models have up to a 33-percent longer maximum flight time (40 minutes) than the Phantom 4 models (see Figure 1.7). Second, the Matrice and Agras MG models can carry payloads of approximately 5,500 g and 15,000 g, respectively (see Figure 1.8).[4] These flight times and payload weights make them especially suited for nefarious activities.

Like other sUASs, DJI's models and their performance characteristics are highly sensitive to large leaps in component and subsystem technology. Chapter Two of this report explores these concepts and how predicted increases in, for example, battery capacity could radically improve the performance values shown here.

Figure 1.7
Model Maximum Flight Times of Various DJI Industrial Models, 2013–2018

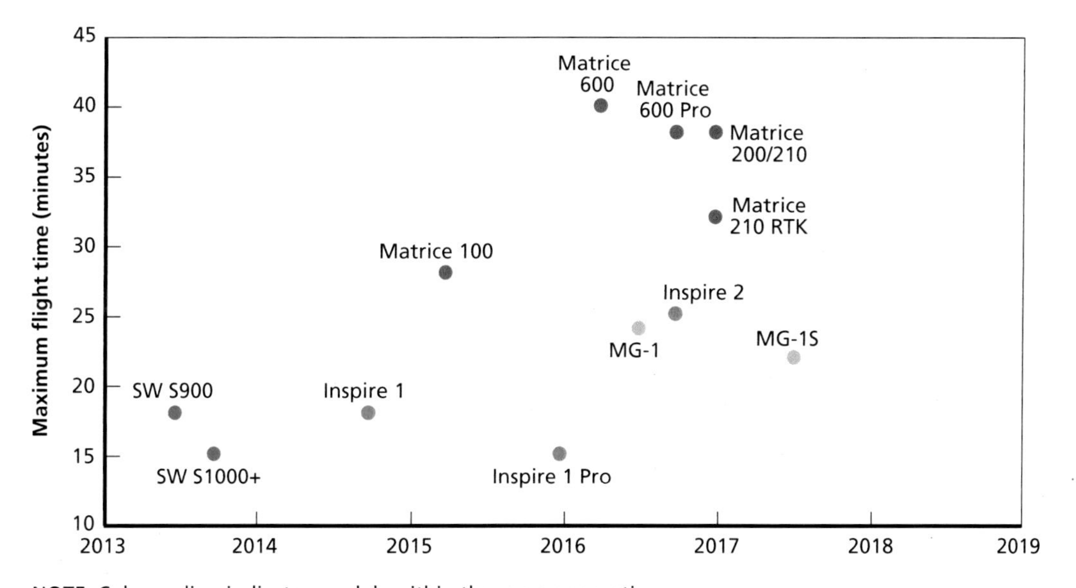

NOTE: Color-coding indicates models within the same generation.

[4] The DJI Agras MG is an agricultural spraying sUAS.

Figure 1.8
Maximum Payload Weight of Various DJI Industrial Models, 2013–2018

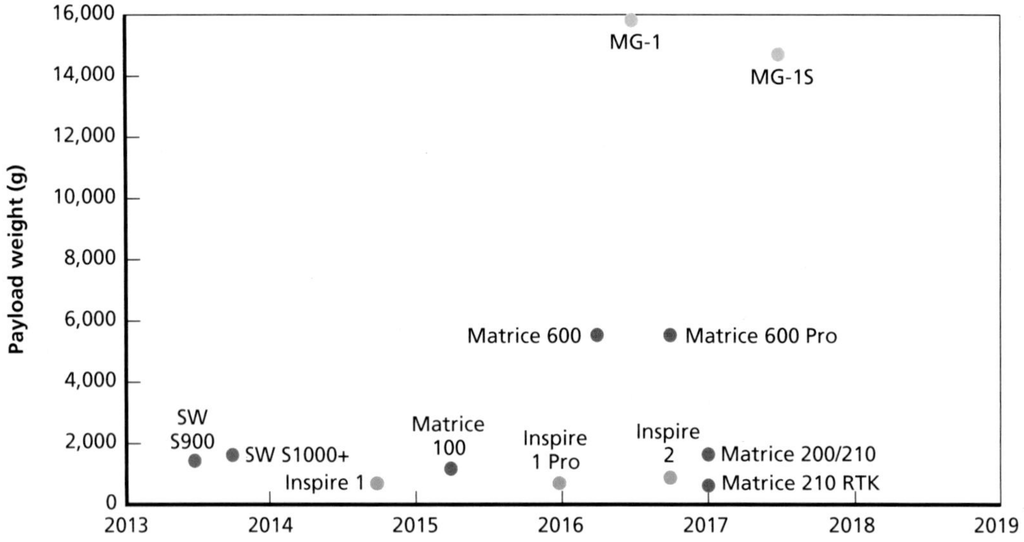

NOTE: Color-coding indicates models of the same platform type.

Organization of This Report

The remainder of this report is organized as follows. Chapter Two examines sUAS technologies across many characteristics and measures. Chapter Three explores the market more broadly and describes our analysis of trade-offs among characteristics and measures. Chapter Four examines how nefarious actors may threaten DHS interests with sUASs. Chapter Five discusses how sensors are used to detect, track, and identify sUASs and their potential utility for C-UAS operations.

CHAPTER TWO

Technology Capability Assessment

This chapter reviews the capabilities of today's sUASs. First, it describes the various attributes of sUASs, including their types, materials, and ways to estimate their payloads. It then describes various payloads and sensors, including nefarious uses of radio frequency (RF) payloads on small UASs. It concludes by describing the C3 attributes of UASs.

Airframes

Lift Generation

Most sUASs are traditional fixed-wing or rotorcraft platforms. Rotorcraft are mostly multicopters. Figures 2.1 and 2.2 show this trend.

Figure 2.1
Airframe Design

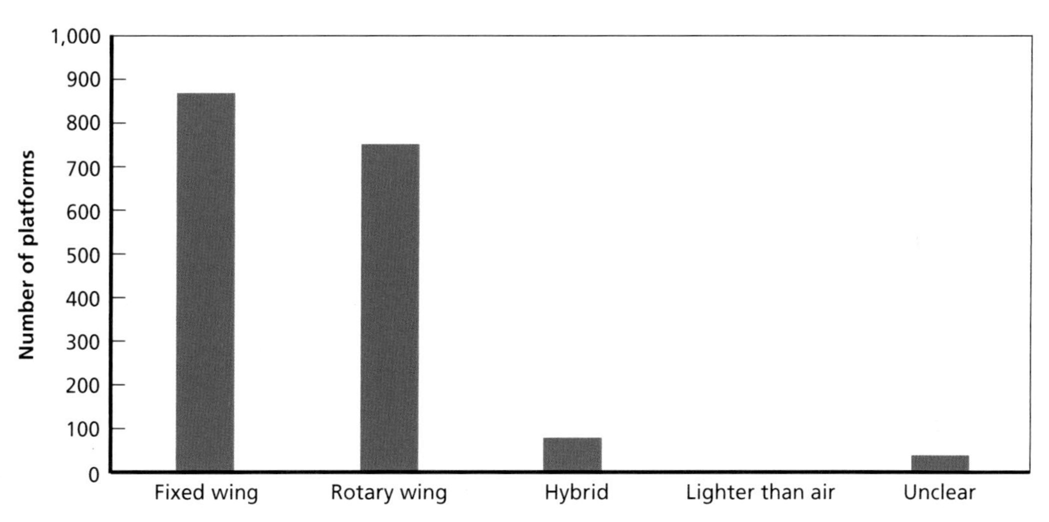

SOURCE: Data drawn from the consolidated database developed for this study.
NOTE: The "Unclear" category includes models on which we had insufficient information to assign a category.

Figure 2.2
Rotorcraft Design

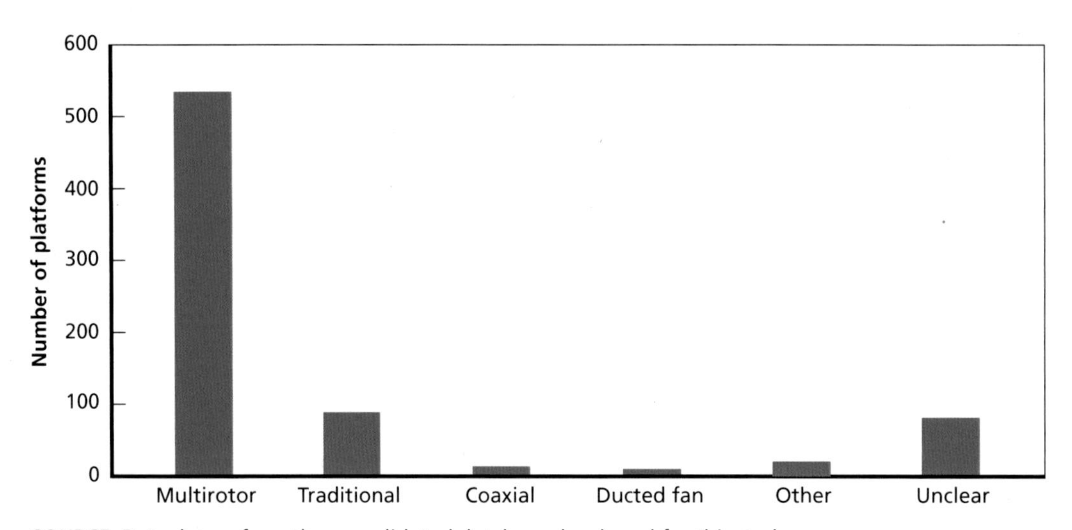

SOURCE: Data drawn from the consolidated database developed for this study.
NOTE: The "Unclear" category includes models on which we had insufficient information to assign a category.

Fixed-wing is the oldest style of UAS chassis. These systems have a standard fixed, two-wing implementation similar to most passenger planes. The wings provide the system with the ability to glide when the motor is disabled. Additionally, the aerodynamics of these systems allow for increased speed and flight time, though at the cost of the multirotor's agility and vertical takeoff and landing (VTOL) capability.

An sUAS chassis designed to hold multiple vertically mounted propellers is commonly called a *multicopter*. These systems have two to eight lift-generating propellers, allowing a large variety of sizes, shapes, and speeds. They are capable of VTOL, hovering in place, and performing complex maneuvers due to the agility and balance provided by the multiple rotors.

A noticeable minority of platforms are hybrid systems, favoring a unique method of lift generation that combines fixed- and rotary-wing elements. Most common of these are quadplanes, which have fixed-wing lifting surfaces and quadrotor multicopter rotors. We describe these unique design choices below, along with their implications:

- *Multirotor and fixed-wing:* A fixed-wing–like airframe that also has rotors, enabling it to achieve VTOL or short takeoff and landing (STOL). During flight, the rotors may either spin freely or be used in an autogyro-like manner to add lift. These platforms are designed to have rotary-wing utility with no runway need and with fixed-wing endurance and range. Quadplanes, combining a quadcopter rotor structure with fixed-wing lifting surfaces, are the most common of these.

- *Powered parachute:* These platforms use a propeller to move while a parachute provides lift. Similar in form to manned paraplanes, they are inexpensive, durable, and easy-to-control platforms that can be packed into very small spaces for transport.
- *Tilt rotor:* Engine nacelles pivot between vertical and horizontal positions offering VTOL/STOL and fixed-wing range and endurance at the cost of added mechanical complexity and weight. A familiar example would be the V-22.
- *Tail sitter:* These fixed-wing aircraft have relatively large, powerful propellers that allow the system to achieve VTOL standing on its tail, then tilt forward into fixed-wing flight once airborne. These models offer fixed-wing range and endurance with a VTOL capability but are limited in design by their requirement for the propeller to lift the aircraft and the large resulting momentum this produces for the airframe.
- *Flapping:* These are very small sUASs that mimic the flight of birds or insects by flapping to stay aloft. There is a potential for energy-efficient flight with this design, but most devices are still research prototypes and are very small, limiting their useful payload.

Finally, lighter-than-air craft make up a very small number of the sUASs examined. These systems often consist of a blimp-like structure and were not common in our market survey, though radio-controlled blimps are already in use and can be converted to sUASs. This type of lift represents a method for attaining very long endurance, though at the cost of speed.

Combining lighter-than-air methods with rotary-wing lifting surfaces offers another avenue for exchanging speed for increased lift in a more limited manner than using lighter-than-air lift only.

Material Trends

Airframes for sUASs are primarily composed of one of a few selected materials. Carbon fiber composites, plastic, aluminum alloys, and foam are the most common. On occasion, balsa wood is used (mainly for trainer fixed-wing sUASs).

The majority (of those identifiable) use some type of carbon fiber composite. Molded plastic, foam, and aluminum alloys are also commonly used primary materials but to a lesser degree than carbon fiber. Figure 2.3 shows this breakdown.

Additive manufacturing, otherwise known as three-dimensional (3D) printing, is a manufacturing rapid prototyping method in which material is joined or solidified under computer control to create a 3D object. A hobbyist can purchase a 3D printer and download or design parts and print them in polylactic acid or acylonitrile butadiene styrene material. Commercial 3D printers are also capable of printing metal and graphene components. There are sUAS plans that can be downloaded from the internet, and more plans become available every day. Parts and components can also be

Figure 2.3
Primary sUAS Airframe Material

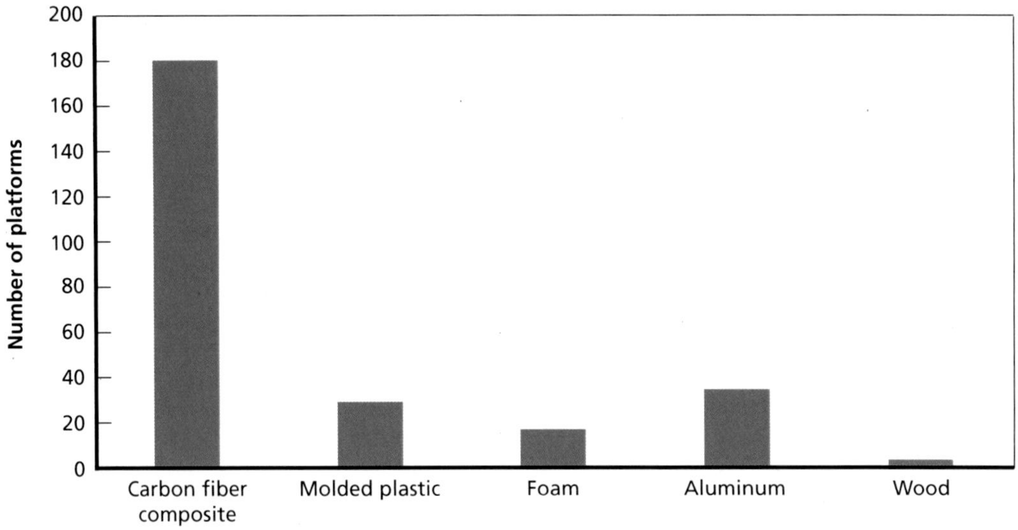

SOURCE: Data drawn from the consolidated database developed for this study.

designed and printed to adapt a commercial off-the-shelf system to carry an external payload.

Material Utility

The previous section showed that carbon fiber composites are the most common choice for sUAS construction. To understand why this is, we looked in the physical characteristics of common current and likely near-future sUAS materials.

Material strength may be measured by many parameters. For airframes, the strength of the material and weight are critical. An ideal material would have a low density and high strength, but real materials often compromise among these or other parameters, such as cost or workability. Some offer clear advantages, as shown in Table 2.1.

Based on the trends in materials used in sUAS manufacturing, we came to the following conclusions:

- Foam and, occasionally, wood are used for extremely lightweight, low-payload sUAS applications when strength is of less concern than inexpensive, easy manufacturing.
- Molded plastics are often used for coverings that offer some protection for delicate parts or visual flair, as well as on structures that do not carry large payloads.

Table 2.1
Airframe Material Physical Properties

Material	Density (g/cm^3)	Ultimate Tensile Strength (MPa)	Shear Strength (MPa)
Balsa wood	0.2	13.5	3.0
Aluminum (6061-T6 alloy)	2.7	310.0	207.0
Polyurethane foam	0.4	23.0	3.7
Molded plastic (acylonitrile butadiene styrene)	1.1	42.0	—
Carbon fiber composite (CYCOM 950-1)	1:4	535.0	87.0
Titanium (Ti-6Al-4V)	4.4	950.0	550.0

SOURCE: Data from Lindeburg, 1995; MatWeb, undated.

NOTE: MPa = megapascals, a measure of material strength.

- Carbon fiber offers a significant increase in strength with minimal weight increase when compared to foams, woods, and plastics. It is ideal for creating strong, lightweight structures.
- Aluminum is a relatively lightweight and strong metal that offers superior shear strength. For this reason, many models use fittings and components made from aluminum.

Steel is occasionally used in aerospace design when extremely strong structures are required, though sUAS fall outside this requirement. Titanium is not used in sUAS construction for similar reasons, as well as cost and ease-of-use concerns.

Graphene and derivative carbon nanotubes offer near-ideal properties for aeronautical applications, incredible strength, and minimal weight, but current methods are unable to reliably produce purified, usable graphene (Lowe, 2018). However, there is reason to believe that production techniques could become commercially viable in the near term (Gorey, 2018) and shift sUAS materials use and construction even more dramatically than composites have done. Pure, useable graphene would dramatically increase the available payload for sUASs by greatly reducing airframe weight.

Looking beyond material selection toward construction, additive manufacturing presents opportunities for explosive growth in design and fabrication. Affordable 3D printers will enable component and even full airframe construction from a range of materials. The ability to rapidly and easily share designs and experimental results online also means that research and development advancement will be driven by the user community. If one person can create something (e.g., a novel design, a stealthy structure) and is willing to share that design, it will be available to all users almost instantly. Designs that are not shared may be copied through 3D imaging.

General Rules for Payload Estimation

Payloads can vary considerably with aircraft design and material selection. However, trends can be observed for common designs and materials. The airframe, propulsion system, and fuel typically account for 60–70 percent of the maximum gross takeoff weight of a rotary-wing sUAS, with the remainder being payload. Plastic airframes trend toward the 70-percent end of that range, while carbon fiber airframes trend toward the 60-percent end. Fixed-wing aircraft trend more toward 70–75 percent. Hybrid airframes, owing to their rarity and multiple design types, make creating such rules more difficult.

Drivers for Future Material Use

Cheap, light consumer sUASs meant for nonprofessionals that carry little or no payload will likely continue to use foam and plastic because these materials are inexpensive, light, and damage-resistant. Professional use will continue to drive carbon fiber for airframes because it provides a particularly attractive combination of high strength and low weight that allows larger sUASs to carry serviceable payloads with the best possible range and endurance.

Payloads and Sensors

Non-Sensing Payloads

Non-sensing payloads are anything the sUAS may carry that does not gather information on the world around it. This can include anything that is able to be lifted by the sUAS, ranging from propaganda leaflets and noisemaking nuisance devices to illegal drugs or weapons—even extremely dangerous biological or radiological weapons.

Non-sensing payloads may be designed to remain with the aircraft at all times or to deploy, possibly during flight. The following is a non-exhaustive list of payload delivery methods that we have observed in our analysis:

- *Kamikaze:* The most rudimentary of payload delivery methods, the payload is strapped to the airframe and the payload and airframe are both crashed into a target. This is, obviously, hazardous to the UAS, and the UAS must be considered expendable when delivering a payload in this way.
- *Payload release:* Systems carry a payload and are able to release it mid-mission. This approach may be used to drop or launch anything from informational pamphlets to explosives or sensors, such as a Wi-Fi sniffer or unmanned ground sensor. These types of payload delivery systems allow the sUAS to be reused.
- *Sprayers:* These systems are designed to spread aerosolized products below the UAS. They are commonly found in agricultural UASs to spread pesticides or fertilizers. A nefarious user could use this type of payload delivery system to spread chemical, biological, or radiological contaminants.

Sensing Payloads

An sUAS can carry many different types of sensors. Indeed, the type is limited only by payload capacity and communication bandwidth. However, the more sensors in use, the greater the energy consumption. Although the most common sensors, by far, are electro-optical (EO) or infrared (IR) cameras and video cameras, many other sensors are used:

- *Still and video cameras:* In many cases, EO video feeds allow pilots to steer an aircraft in first-person view, in addition to providing other surveillance functions. On an sUAS, EO cameras are capable of recording in 4K resolution, creating recordings with great clarity and high detail, though communication bandwidth limits real-time feeds to 2K resolution. Their use for surveillance and reconnaissance is also well documented. Optical flow, the apparent motion of objects, surfaces, and edges in a visual scene caused by the relative motion between an observer and the scene, can be achieved with these cameras. Physical limitations, such as optical diffraction, can limit lens sizes, but current cameras are already small enough to allow additional payload in some cases. IR cameras are less common due to their limited utility for first-person vision, but their use in agricultural monitoring and target tracking for "follow-me" applications means that they are also readily available, especially on larger sUASs.[1] IR images can also reduce the required communication bandwidth.
- *Hyperspectral:* Commercial research is now enabling hyperspectral imaging devices to be miniaturized for sUAS use (Hinnrichs, Hinnrichs, and McCutchen, 2017). These sensors are useful for agricultural and geoscience applications, so their continued development and expansion to commercial UASs is highly likely.
- *Light detection and ranging (LIDAR):* This phenomenology uses pulsed laser illumination and sensitive electronics to measure the time needed for light to reflect off a target and return to the LIDAR device. LIDAR thus enables highly accurate 3D depth estimates based on knowledge of the speed of light. Miniaturization of LIDAR systems geared toward deployment on UASs has resulted in a variety of commercial applications for LIDAR payloads.
- *Radar:* Radar miniaturization is near or at levels conducive to sUAS use. Radar technology firm Echodyne has created radar units that weigh less than a kilogram and can be carried by multirotor sUAS (Echodyne, undated). Academic research has shown that combined payloads that include a radar element can be as small as 22 lbs (Huizing et al., 2009). In the Echodyne case, the radar was limited to a 3-km range but weighed only 1.65 lbs and consumed 30 watts. Continued work in this area, driven by regulations requiring radar navigation of aircraft in certain

[1] Follow-me mode allows a UAS to hover over a user or target and is often used to capture video of athletes.

situations, is likely to miniaturize these units even more. The size and weight may soon be suitable for sUAS applications.

- *Electronic intelligence:* UASs offer an ideal platform to catalog electronic signals due to their freedom of movement and agility. This may come in the form of passive collection, such as wardriving,[2] active interception of signals, or even nefarious disruption or spoofing of electronic signals.
- *Acoustic sensors:* These sensors have been proposed for use in navigation and use biomimicry to echolocate nearby obstacles. They have also been proposed for atmospheric tomography, and there is the potential for acoustic identification of other UASs. These uses are still largely theoretical (Kapoor et al., 2017).
- *Altimeters:* An altimeter is an instrument used to measure the altitude of an object above a fixed level. In the case of sUASs, a sonar or laser altimeter can be used to aid in measuring the distance from the ground for take-off, landing at very low altitudes, and nap-of-the-earth flight to avoid detection systems.
- *Radiation:* Essentially a flying Geiger counter, UASs with radiation sensors have been used to detect gamma, X-ray, alpha, and beta particle radiation.

Sensors as Navigation Aids

In addition to collecting ISR or targeting information, data sensors can be used as navigation aids. A familiar application is global navigation satellite system (GNSS) use. Additionally, the inertial measurement unit onboard an sUAS can be used as part of an inertial navigation system (INS). These systems can measure changes in position and velocity by integrating the variations in acceleration over time. This integration does mean that small errors are compounded by the integration, and the calculated position and velocity slowly drift away from their true value unless they are updated with new, accurate initial conditions (usually via GNSS). But on small time and distance scales, this method can be very effective. For example, an sUAS that is trying to avoid spoofing interference during its terminal attack phase can shut off its GNSS receiver.

Visual sensors can also be used as navigation aids. Optical flow measurements allow a sensor to measure the relative rate of movement between the sensor and the scene it is observing. If known fixed targets are in the scene, they can be used as reference points to calculate positional measurements.

Size Limitations of Electro-Optical Sensors

EO sensor sizes are limited by the ability to discernably differentiate between point light sources due to the diffraction of those light sources as they are sampled by the sensor (Cambridge in Colour, undated). The Rayleigh Criterion provides a convenient method for defining a break point where differentiation between point sources

[2] Moving around a physical space and cataloging Wi-Fi networks, ostensibly for future exploit.

is no longer sufficient and begins to affect image quality and detail (Nave, 2016b). Equation 2.1 shows how to calculate this value for a circular aperture:

$$\sin\theta_R = 1.22\frac{\lambda}{d}, \tag{2.1}$$

where θ_R = angle of the first diffraction minimum, λ = wavelength of the light source, and d = diameter of the sensing aperture.

When the angle of the first diffraction minimum approaches 90 degrees, the image becomes unresolvable (Nave, 2016a). With this information, we can show that effective pixel sizes may not get lower than approximately the wavelength of light being sampled.

With EO sensors sampling light sources from 10^{-5} to 10^{0} millimeters, pixel (and correspondingly sensor) size can vary significantly with the type of light being captured. But this value provides a rough estimate of sensor size. For example, a sensor sampling visible light (~500-nanometer wavelength) would need pixels with an effective size of at least 600 nanometers. If the associated sensor was seeking a 1,000 × 1,000–pixel image resolution, the sensor would need to be around half a millimeter in width to obtain a useful image under ideal conditions. Because of wavelength, IR cameras would require larger sensors, while ultraviolet cameras would require smaller sensors.

Nefarious Use of RF Electronic Payloads on Small UASs

The use of sUASs as a method to counter nefarious sUASs is being examined as a potential option to combat this ever-changing threat. However, RF electronic payloads on adversary sUASs are a viable threat to these types of countermeasures (in addition to their normal nefarious uses), and the capabilities of such systems should be examined. Examples include Global Positioning System (GPS) jammers, GNSS spoofers, mobile phone jammers, and the collection of air traffic control signals. Additionally, several technical trends could affect RF payloads on sUASs.

GNSS Jammers

One possible threat posed by nefarious actors involves electronic payloads on sUASs that include GNSS jammers or spoofers. These devices weigh less than a pound and are under eight inches on their long axis—meaning they can be carried by many sUASs. Jammers would render GNSS receivers near the sUAS inoperable by significantly reducing the signal-to-noise ratio for the GNSS receiver channel, thereby disabling or disorienting navigational systems that rely on GNSS.

GNSS signals are broadcast on Earth from satellites orbiting at medium Earth orbit. The signals that impinge on the Earth's surface are weak, on the order of –136 dBW/m^2. When received by an antenna with zero gain, the power reaching a GNSS receiver would be –130 dBW. One nefarious application is to emit enough RF energy in the GNSS bandwidth to overwhelm the GNSS signal to GNSS receivers in the vicinity of the jammer. Although they are illegal to operate in the United States, even by police, very small jammers a few inches long and weighing a few ounces can be bought on the internet (Hill, 2017; Jammer-Store, undated). One common system, the GJ 6, weighs 2 lbs and includes a battery that can operate for three hours, consumes 10 W, and a has range of 20 m.

However, the danger of a GNSS jammer as a payload on an sUAS is the ability to affect GNSS receivers over a wider area than a ground-based jammer, owing to the high altitude and resultant longer lines of sight achievable with an sUAS. The range of GNSS jamming depends on power. The quoted range of 20 m for the GJ6 is probably conservative because the estimate accounts for physical obstructions between the jammer and the GNSS receiver. However, with an unobstructed view from an sUAS flying at a high altitude, all that would be required is adequate RF power emitted from the jammer to overwhelm the GNSS signal over a very wide area.

Equation 2.2 shows the required RF output power (*J*) for a jammer to deliver a certain jammer-to-signal ratio (*J*/*S*) at a GNSS receiver (*r*) distance away:

$$J = \left(\frac{J}{S}\right)\frac{4\pi r^2 P_r}{G}, \tag{2.2}$$

where $P_r = -130$ dBW is the GNSS signal power received by the GNSS receiver, and $G = 0$ dB is the assumed gain of the GNSS receiver antenna.

Figure 2.4 shows the required RF jamming power in milliwatts as a function of distance *r* for several values of *J*/*S*.

Figure 2.4
Required GNSS Jam Power Versus Unobstructed Range and J/S

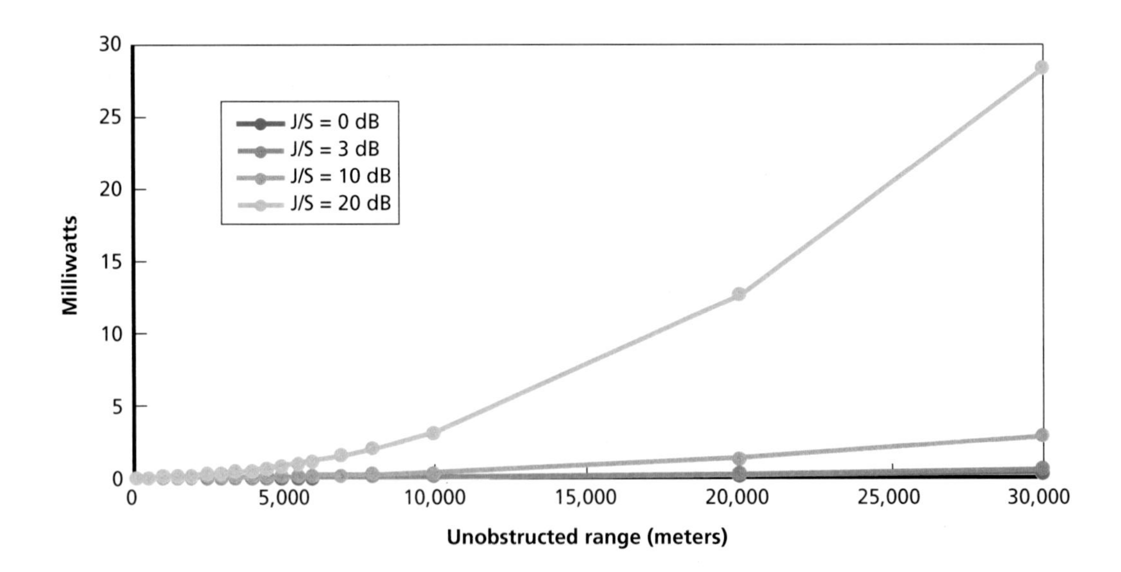

Figure 2.4 shows that, due to the weak GNSS signals received on Earth, less than a watt of RF power is required to jam GNSS receivers many kilometers away, as long as the range is unobstructed. Hence, a 1 W GNSS jammer on a high-flying sUAS has the potential to jam GNSS receivers with 20 dB of J/S tens of kilometers away.

Often, when considering GNSS jamming on sUASs, the expectation of the sUAS returning to home is a consideration. An sUAS cannot return home when it does not know its current location. The sUAS will remain in place and hover when the GNSS locating system is disrupted. Additionally, fixed-wing sUASs may no longer have the capability to fly waypoints to a target; however, forward motion is not halted. Finally, alternative methods of navigation, including internal maps, optical flow, internal navigation units, and inertial measurement units, can aid in GNSS-denied navigation. In these cases, a counter system may prove less effective if these responses are not considered when determining a protection strategy.

GNSS Spoofers

The same range equations for GNSS jamming apply to GNSS spoofing. In fact, GNSS spoofing may require even less range than jamming, since the goal is to fool the receiver instead of overwhelming the legitimate GNSS signal with jam power. The easiest way to spoof a GNSS receiver is to deliver a legitimate GNSS signal but delayed in time from the real signal. This can be achieved with digital RF memory (DRFM) techniques whereby GNSS signals are received, digitally copied, and rebroadcast verbatim by the spoofing transmitter. Because GNSS works by comparing the time of arrival of signals from different satellites, rebroadcasting a legitimate GNSS signal at the wrong time will cause the GNSS receiver to produce an erroneous location.

Mobile Phone Jammers

The same range equations also apply for mobile phone jammers, except that the maximum received signal power of a cell phone with electronic support measures (ESM), P_r, is about –45 dBW (–15 dBm), more than eight orders of magnitude larger than a nominal GNSS signal. The jamming range equation applies, except for the higher received signal power P_r.

As Figure 2.5 shows, the required RF power to jam cell phones is much higher (measured in the watts versus milliwatts) than for GNSS, since the cell jammer must compete with signals from more powerful cell towers that are only a few kilometers away. To achieve a 1-km jam radius, the jammer must emit approximately 400 W of RF power output, which is a hefty amount for an sUAS. One way to reduce power consumption is to use duty factoring—that is, to burst the maximum RF power on for a short duration, just long enough to interfere with the reception of the cell phone, then turn it off. Thus, assuming an amplifier efficiency of 75 percent and a duty factor of 10 percent, 400 W peak RF power can be transmitted to interfere with cell phones

Figure 2.5
Cell RF Jam Power Required

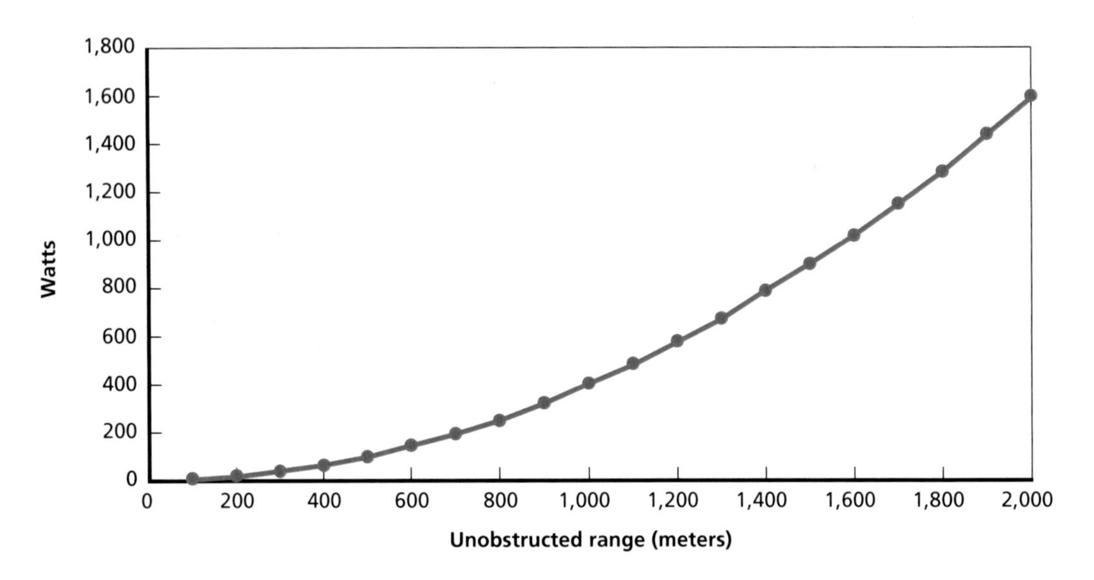

up to 1 km away while consuming only 400 W × 10 percent/75 percent, or 53 W, of DC power.[3]

Air Traffic Control Signal Collection

One potential risk factor lies in the ability of sUASs to use an automatic dependent surveillance-broadcast (ADS-B) air traffic control receiver to track incoming civilian aircraft flights within line of the sight of the receiver. ADS-B is a new standard in U.S. air traffic control whereby civilian aircraft periodically broadcast their GNSS location and flight velocity vectors with an ADS-B transmitter. Any ADS-B receiver within line of sight of the transmitter can receive such information and map the aircraft's location and direction, potentially allowing a collection range of hundreds of miles.

One potential danger that could arise if nefarious actors connected an ADS-B receiver to an sUAS navigation system is that they could guide the sUAS into the flight path of incoming aircraft. A University of Denver study noted that an sUAS weighing only 2 kg striking a business jet flying at cruising altitude would deliver the same amount of energy (54 kJ) as a 20-mm anti-aircraft cannon shell (Moses et al., undated). Although impact with an sUAS would not have the equivalent ballistic effect of a cannon shell, this relatively large amount of energy still presents a lethal risk if the sUAS can be guided into the path of jets flying at high cruising speeds with the use of an ADS-B receiver. ADS-B receivers are compact, widely available, and inexpensive,

---

[3] Representative of a typical class AB amplifier efficiency using the Empower RF model BBM3Q6AHM as a surrogate (Empower RF Systems, Inc., 2016). Assuming an attack on a 5G physical broadcast channel, a low-duty cycle, such as 10 percent, is plausible to prevent a mobile device from connecting to a cell (Lichtman et al., 2018).

making this a feasible risk. Amazon sells one that is three inches long, weighs only four ounces, and requires a small lithium-ion (Li-ion) battery for power—small enough to fit on most sUASs (Amazon.com, undated).

Technical Trends Affecting sUAS Payloads

Aside from the well-known Moore's law,[4] we highlight four additional technical trends that could affect the proliferation of RF payloads on sUASs.

Analog and Digital Converters

The first trend is the increasing bandwidth of mixed-signal devices (e.g., analog-to-digital [A/D] and digital-to-analog [D/A] converters). The so-called Walden curve, for example, showed a tenfold increase in bandwidth of A/D converters over ten years, controlling for resolution, as shown in Figure 2.6.[5]

Similar improvements in the bandwidth (speed) of D/A converters have been seen as well. Increases in the bandwidth of A/D converters allow compact RF payloads to digitize wider bandwidths of RF signals with better resolution and to digitally process

Figure 2.6
Walden Curve Showing a Tenfold Increase in A/D Speed Over Ten Years

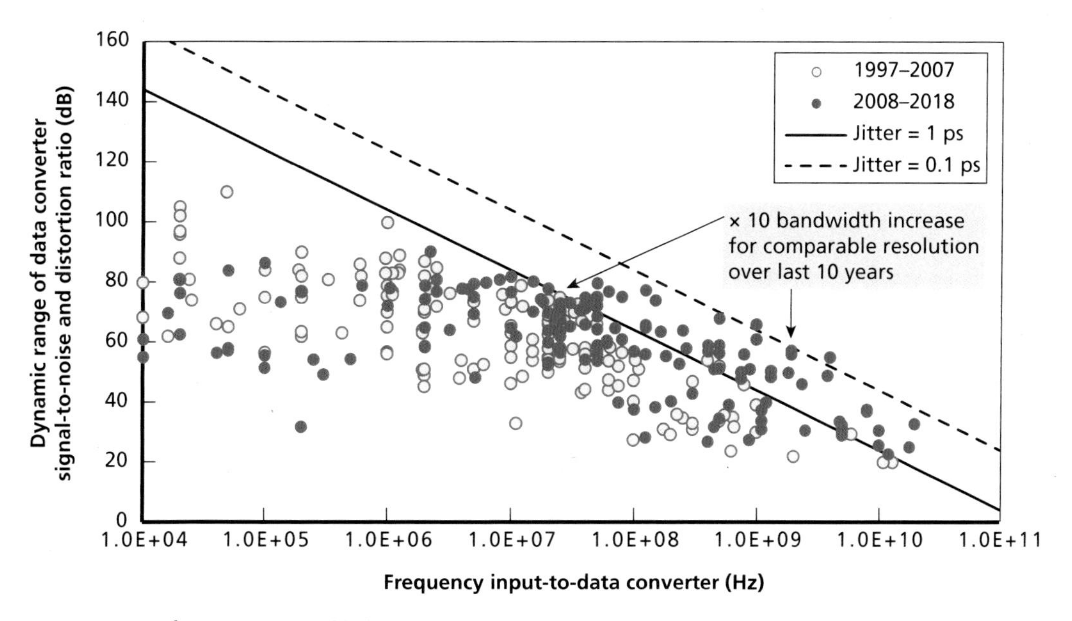

SOURCE: Data from Murmann, 2018.

---

[4] Gordon Moore observed and predicted the doubling of the components that can be squeezed into electronic circuitry and the halving of cost in the 1960s, a prediction that held true until recently.

[5] The Walden curve describes the performance of systems that convert analog signals into digital signals, such as those used as UAS sensors.

larger chunks of the RF spectrum. The increase in D/A converter bandwidth, reciprocally, provides the ability to generate and transmit wider bandwidths of RF signals, as well as a greater diversity of signals, or to generate and transmit signals across multiple and disparate bands. Increasingly powerful mixed-signal technology allows for significant size, weight, and power (SWaP) reductions, as well as better performance of RF circuitry. Higher frequencies and larger bandwidths of RF signals, which require high sampling rates as dictated by the Nyquist-Shannon theorem, can now be digitized directly with a single device. This characteristic allows RF designers to bypass up/down conversion stages of the RF circuitry that traditionally increased weight, power, and heat in earlier generations of RF systems.

Field Programmable Gate Arrays

The second trend we examined highlights improvements in field-programmable gate arrays (FPGAs) that enable high-speed processing of RF signals in a programmable firmware environment. FPGA programming enables massive parallelism so that larger bandwidths of signal data can be processed at high speeds, and it allows the same hardware to handle multiple RF modes and functions simultaneously. The footprint (number of gates) of commercially available FPGAs is increasing, as are clock speeds. This feature allows more computations to be performed within a single FPGA, reducing the need for multiple FPGAs, as well as size and weight reductions. This is an advantage when performing the massively parallel, high-speed signal-processing computation required by multifunction RF payloads. Of course, power draw and heat generation may be a concern when increasing the onboard computation load, something manufacturers will need to be cognizant of and address with some combination of adding throttling computation capacity, adding cooling capability, or increasing energy capacity.

This increase in FPGA capability directly ties into greater capability for software-defined radios (SDRs) as well. SDRs have replaced common hardware portions of radio transmitters, such as amplifiers, filters, and mixers, with software that performs the same function. Removing hardware reduces weight, of course, increasing the amount of energy available for payload, range, or endurance. But it also places a greater strain on the computational resources of the UAS. The increase in FPGA capability described here means that the added SDR computational load can be handled with little or no increase in processing power draw or equipment weight. The net result will be a reduction in SWaP demands for the radio components of the aircraft. In addition, SDRs offer capabilities beyond the means of hardware-based radios:

- Spread-spectrum technology and software-defined antennas allow the greater use of narrow frequency bands with little interference by using digital signal-processing techniques to correct errors caused by existing interference.
- Cognitive radios allow competing users, such as multiple sUASs in a given geographic area, to communicate and cooperatively optimize spectrum use, again

reducing interference by dynamically allocating portions of the frequency bands in use.

- Dynamic power adjustment on transmitters based on information sent from receivers can increase battery life because of their lower power draw in quiet environments. And they can do so while maintaining the ability to "shout" over noisy areas when using the standard power draw.

Digital Antenna Array Architectures

Generally improving digital technologies, including both central processing units (CPUs) and FPGAs, are also ushering in the third trend of more powerful digital antenna array architectures. These architectures enable sophisticated digital beamforming at high speeds without the cumbersome, dissipation-prone, and expensive analog technology of previous generations. Digital beamforming allows antenna performance to be tailored across disparate RF functions or missions rapidly or simultaneously. Beams can be shaped or steered without mechanical parts or gimbals. New system-on-chip technology also enables multiple waveforms or bands to be transmitted or received simultaneously through these beams with a single set of hardware. With these three technical trends, more-capable multifunction RF payloads can be developed at a fraction of the SWaP of older generations of RF systems.

A technical team at TNO Defence, a public company in the Netherlands, has developed a so-called scalable multifunction RF (SMRF) payload system for sUASs that performs multiple RF functions across multiple beams, pulse repetition frequencies (PRFs), sensitivity levels, and bandwidths, and radiates 2 W of power (Huizing et al., 2009). Table 2.2 summarizes the SMRF modes.

Table 2.2
Nonexhaustive List of RF Modes That Can Be Performed by SMRF Technology

	Characteristics		
Modes	**PRF**	**Sensitivity**	**Bandwidth**
Synthetic aperture radar (SAR)	Medium	Medium	High
Ground moving-target indicator (GMTI)	High	Medium	Low
Sea surveillance	Low	Medium/low	Medium
Sense and avoid	High	Medium	Medium
Communication	N/A	Medium	Medium
Navigation	Low	Low	Low
Weather radar	Low	Medium	Low
Weapon location	High	Very high	Medium
Radar ESM	N/A	Medium	High

High-speed mixed-signal technology allows RF functions using multiple bands to be interleaved in almost real time. FPGAs perform the high-speed signal processing functions in parallel, allowing several RF modes to be used near-simultaneously at high speeds. Firmware programming of the FPGA allows signal algorithms to control the PRF, sensitivity, and bandwidth of RF modes in real time. Radar ranges (SAR, moving target indication) are expected to be in the several kilometers. ESM functionality includes the ability to detect and geolocate via angle-of-arrival information emissions in the X-band region.

We highlight the SMRF example to illustrate how advances in commercial technology are facilitating the development of highly capable, multifunction RF payloads small enough to fit on sUASs. Numerous RF modes, such as SAR, GMTI, maritime surveillance, sense and obstacle avoidance, data communication, navigation, and weapon location (projectile detection and tracking), are possible and can be packaged in a single, compact set of hardware on an sUAS. Tracking of slow and intermittent movers (such as people) has been driven by the capability of space-time adaptive processing. By enabling more computational power with smaller SWaP footprints, advancements in commercial CPU and FPGA technology, along with their compact integration with mixed-signal technology, will facilitate the development of multifunctional RF payloads at a lower cost.[6]

Nefarious uses of multifunction RF payloads could include the following:

- all-weather/night imaging of infrastructure for intelligence gathering on potential targets of opportunity
- tracking of air or ground moving targets, such as vehicles, aircraft, rotorcraft, or UASs
- tracking of maritime vessels
- signals collection to gain situation awareness of the spectral environment
- electronic surveillance to determine vulnerabilities in communication infrastructure.

Graphical Processing Units

An additional technical trend affecting sUAS payloads is the development of graphical processing units (GPUs). GPUs have rapidly changed how large-scale intensive computations are performed. The current trend of using GPUs to perform heavy computational tasks previously handled by CPUs has offered an incredible increase in overall system speed while being transparent to the end user. While CPUs include fast-cache memories and are still valuable for sequential and logical tasks, GPUs have a different

---

[6] Ettus Research offers a powerful, fully programmable, software-defined radio platform for less than $5,000 that integrates FPGA, mixed-signal, and microprocessor technologies in a compact form factor. Models cover DC-6 GHz with 160 MHz base bandwidth and include high-speed interfaces, the large user-programmable Kintex-7 FPGA, and a PowerPC core, potentially supporting many of the RF functions listed by TNO Defence.

structure, with hundreds or thousands of different threads (usually used to render an image), and their functionality offers high throughput for parallel tasks. The architecture of GPUs shows opens computational power that was not previously available, allowing the sUAS to use more-complex algorithms. With this capability, sUAS systems that have recently appeared on the market are capable of performing flight control, obstacle avoidance, and target recognition and pursuit simultaneously. A 2012 conference paper by Harju et. al. provided an initial graphical example of the differences in computational power between CPUs and GPUs using data from 2001 to 2012 on floating point operations counts (Harju et al., 2012). Recent updates of these data include GPU systems introduced since 2014 and illustrate the dramatic increases in this technology's ability to perform complex calculations with low size, weight, and power requirements.

Figure 2.7 shows the CPUs from Intel, some of the most powerful CPUs available, as blue lines. The GPUs developed by NVIDIA and sold on the consumer market

Figure 2.7
CPU and GPU Theoretical Peaks

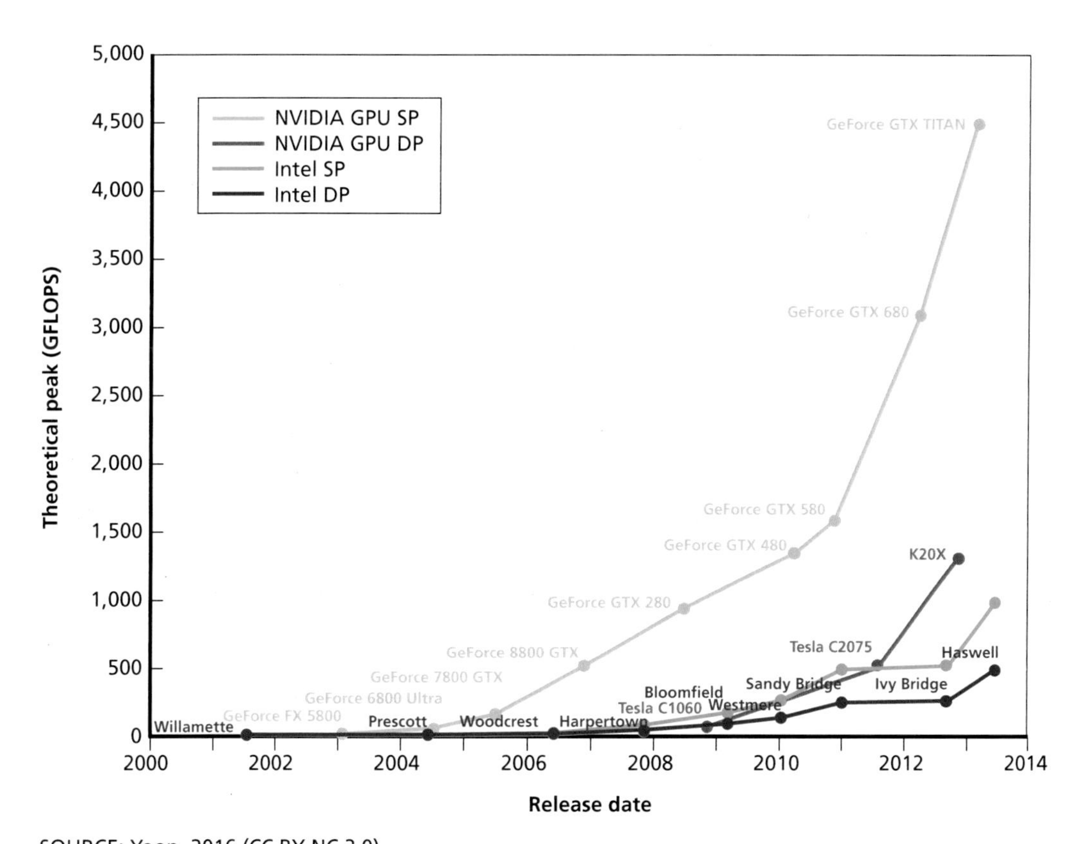

SOURCE: Yoon, 2016 (CC BY-NC 3.0).
NOTE: Floating point operations per second (FLOPS) are used to measure computer performance; the scale in the figure uses gigaFLOPS (GFLOPS).

for desktop and laptop computing systems are shown in green. At the time the Titan GPU was released, it was capable of 4.5 times as many calculations per second as the best CPU available.

The 2018 release of the NVIDIA Tesla V100 shows this trend of rapid growth continuing, with a theoretical peak of over 7,000 GFLOPS (Xcelerit, 2018). The result of this trend is smaller, lighter processing power with a lower power draw, enabling "smarter" sUASs.

A/D converter improvements will have a large impact on RF payloads, but these improvements will also have a significant impact on sUASs themselves. A/D conversion permeates every component of an sUAS. All commands from the sUAS to its motors involve translating a digital command to analog voltages. All sensor information provided to the sUAS is collected as analog information and is transformed into digital information. These base components of sUAS navigation, behavior, and use are all directly controlled by the speed of A/D converters. Increased speeds will increase performance in all aspects of the sUAS. This trend can be observed in the rapidly developing area of flight controllers for racing quadcopters. Newer processors allow greater control and more features to be implemented onboard.

Command, Control, and Communications

C3 capabilities are critical to UAS mission success. Without proper command and control, whether from an operator or from autonomous systems, a UAS cannot perform the actions to carry out its mission. Similarly, broken or sufficiently degraded communication leaves the UAS without operator input, unable to transmit information back to the operator, and, potentially, without guidance. These situations are obviously undesirable for both legal and nefarious operators, meaning the industry is likely to push toward robust C3 solutions that may make it difficult for DHS to defend against sUAS threats.

Command and Control

Autonomy

The ability of an sUAS to perform path-following, path planning, follow-me, obstacle avoidance, small-team coordination, and swarm formations has been proven in a variety of contexts.[7] Emerging from the theoretical world of modeling and being observed as plausible and potential force multiplier is the capability of drones to flock. How-

[7] Swarm is a loosely defined term generally related to autonomy and group coordination. While it is possible to differentiate among multiple levels of such tactics (e.g., level of coordination, communication among members, number of members), in this report, we define a swarm as a group of sUASs that are in active communication with each member of the swarm, adapting their behavior based on the behavior of other members of the swarm. This is in contrast to a multiagent team, which has coordinated members that operate in a preplanned manner.

ever, the autonomous capability of the individual drones varies significantly. Some drones are completely dependent on human input for all aspects of flight; such systems are often referred to as being *remotely piloted* and have little autonomous capability.[8] Others require only a few key human inputs and can operate in an intelligent collaborative group with other drones. In this section, we provide an overview of capabilities along the spectrum of autonomous control.

A precursor to autonomy is automation. Many drones have a capability referred to as *path following*. The user programs the flight path for the drone, and each part of the journey is accounted for on a solid path. Beyond path following, the level of automation blends into autonomy, with path planning and follow-me modes often referred to as *self-flying* or *autopilot*. Drones capable of path planning sometimes still require including key checkpoints along the way, or waypoints. Waypoints are often where an adjustment to altitude, direction, or speed occurs as the drone follows a path, connect-the-dots style, that the human user has planned.

Drones capable of follow-me mode are programmed to use sensors to lock on to a particular object and follow its movements. This can range from a device that must be placed on the object or person to be tracked (a remote-control bracelet or a phone that sends out a GNSS signal that the drone follows) to drones with high-quality visual sensors that can track whatever person or object has been selected in the user interface. This functionality leaves the operator unencumbered by the need to control the flight path. Increasing the level of autonomy, the capabilities venture into systems that can be taught rules and make independent flight decisions without direct human input. Taking the path-planning and follow-me capabilities a step further, some models have obstacle avoidance, with the drone's sensors and programming allowing it to detect an obstacle and (without human input) change its flight plan to avoid the unexpected obstacle and then return to its original path.

Drone swarms can initially appear to be highly autonomous. However, these complex patterns and paths are preprogramed by a human (including obstacle avoidance or course correction); they are essentially a form of very complex automated path following and not indicative of adaptive autonomous control. As Andrew Masterson, writing for *Cosmos Magazine*, points out,

> So far, so elegant, but while each of a thousand flying starlings, for instance, is perfectly capable of flying around a tree, a dozen autonomous drones confronted by the same obstacle are likely to crash into it, collide with each other, or fly off in several different directions. (Masterson, 2018)

Simply put: If a group of sUASs does not change its behavior in response to a stimulus (which may include the behavior of group members), it is not a swarm.

[8] These remotely piloted drones sometimes have the ability to hold a position, self-control their landing in an emergency, or perform other tasks with limited autonomy.

Flocking, on the other hand, takes the idea of coordinated flight and increases the level of autonomy, with each individual system knowing three rules that are observed in natural flocks: Do not hit obstacles, do not hit each other, and head toward the same target. These flocks can begin as a group or can self-organize if the control systems include machine learning (Vásárhelyi et al., 2018).

Operator Control

The drone controller market is shifting away from smartphones and tablets due to a lack of precision control and the difficulty of mastering the technology. Instead of a smartphone with a lagging reaction time, which can make dynamic precision adjustments to flight patterns difficult, many advanced drones are starting to incorporate "copilot" technology. This technology involves either a separate control system (dual operator) with a dedicated remote with precision controls or software, allowing the drone to perform semi- or fully autonomous flight. Some areas of commercial application that require skilled professionals with artistic or dynamic inflight decisionmaking (such as filming) are heading toward the separate controls. However, the rest of the market is focusing on increasing the autonomy of the flight software using artificial intelligence (AI), predictive analytics, and computer vision to remove the requirement of a highly trained operator.

Swarming technology is now making its way into the commercial market, as illustrated by the introduction of SPH Engineering's Universal Ground Control Software (UgCS) drone-show platform, which allows the user to control multiple sUASs at once from a base station (UgCS, undated). This is still a step from autonomous swarming, controlled within the swarm, but the U.S. Navy's Perdix swarm demonstration and the Defense Advanced Research Projects Agency's (DARPA's) work with Raytheon on Collaborative Operations in Denied Environment (CODE), among other projects, show that this type of control and UAS behavior are achievable. Swarming sUASs can achieve through coordination what an individual sUAS could not: attacking or surveilling a target from multiple angles, carrying a large payload in piecemeal fashion (such as a plastic explosive or weapon parts and components), or physically denying a region of flight or observation. Heavy academic and commercial interest in similar concepts suggests that swarms of sUASs are a commercial possibility in the next five to ten years, and small-team capabilities are likely to appear even sooner (Gibbons-Neff, 2017; Giangreco, 2018). Beyond programming for autonomous or vaguely guided flight, alternative control methods, such as facial and movement recognition and eye tracking, will likely be demonstrated in the near future.

An article in *The Economist* summed up modern control methodology:

> Military and consumer drones alike are being transformed by rapid progress in two cutting-edge areas of drone research: autonomy and swarming. If you automate away the need for a skilled operator, drones suddenly become much more useful. Military ones that do not require the oversight of a human operator can

> be radio silent and stealthier. Consumer ones can follow runners, skiers or cyclists and film them from above. Commercial ones can fly a specific, pre-planned path over a field, building site or quarry, avoiding obstacles as they gather data. Improved flight-control algorithms, more on-board processing power and progress in machine vision will allow drones to handle more decisions themselves, rather than relying on fallible or inexpert humans. Most existing drones simply move the pilot from the vehicle to the ground. The next generation of drones will not need pilots at all—just orders. ("Drone Technology Has Made Huge Strides," 2017)

In 2008, the theory that one person could control multiple flights of UASs was being tested with novel control models and algorithms. According to research presented at AUVSI in 2008, "The suggested hierarchical, pseudo-fractal control algorithm simplifies command and control of multiple, heterogeneous, autonomous vehicles. With this structure, a single operator or small team could control multiple UAVs effectively" (Darrah, Raj, and Drakunov, 2008). That model and user interface simulation were tested for combat use with a focus on demonstrating that, once the flight controls were autonomous, the operator would focus on payload controls, calling up SAR images when in range of a potential target, analyzing the image, and deciding whether to engage the target. Additionally, the operator could redirect a single UAS to disengage from the current group and integrate with another group. This capability allowed the operator to balance assets as needed if the armed payloads of one group were depleted.

Eight years later, the question has shifted from finding the maximum number of UASs a single operator could control to how many missions a team could perform efficiently:

> It was found that an experienced operator can supervise up to 15 UASs efficiently using moderate levels of automation, and control (mission and payload management) up to three systems. Once this limit was reached, a single operator's performance was compared to a team controlling the same number of systems. In general, teams led to better performances. Hence, shifting design efforts toward developing tools that support teamwork environments of multiple operators with multiple UASs (MOMU). (Porat et al., 2016)

With operators being asked to complete multiple tasks while maintaining control over each individual UAS, the efficiency and thoroughness of task completion dropped as the number of UASs increased. However, when acting in teams, efficiency, thoroughness, and accuracy increased, but the rate of repeat observations also increased, and coordination of efforts became a challenge (Porat et al., 2016). Other research focusing on software-enabled, coordinated small UASs suggests that 40 is the maximum number of UASs a single operator can control. However, for swarms, ongoing research suggests that 40–100 or even thousands could be possible (National Academies of Sciences, Engineering, and Medicine, 2018).[9]

This research is reflected in the civilian market as well with the separation of flight controls from payload controls and either replacing the flight controller with autonomous programming or replacing the operator's physical control with a "copilot." The single-operator-control focus has enabled growth in capacity, and we see swarms used for everything from military exercises to Super Bowl halftime shows. These swarms are homogenous and have a single preplanned task. As the controls and algorithms advance to support coordination between controllers and between systems, the homogenous nature of the swarm could fade and accept multiple types of UASs with heterogenous types of payloads that can function in concert, be operated by a single person or small team, and carry out a variety of tasks efficiently. Building in redundancies would remove single points of failure in the flight and task plan. DARPA has been working on moving the agility aspect forward with its Converged Collaborative Elements for RF Task Operations (CONCERTO) program. "This kind of multifunction system would let smaller UAS swap missions mid-flight, removing the need to land or have multiple craft in the air simultaneously" (Irving, 2017). As Darrah et al. described it, rather than an operator acting as a remote pilot for one or a few vehicles, this allows the controller to act as a mission commander for a squadron, executing high-level command over a diverse group of capabilities and payloads (Darrah, Raj, and Drakunov, 2008).

This cooperation and coordination theory has existed for many years, but in 2018 it was physically demonstrated for the first time that integrating machine learning could further advance the autonomy of the group. A *Wired* article summarized this recent proof of the long theorized autonomous capability of drones:

> These drones have self-organized into a coherent swarm, flying in synchrony without colliding, and—this is the impressive bit—without a central control unit telling them what to do, each of these 30 drones is tracking its own position, its own velocity, and simultaneously sharing that information with other members of the flock. There is no leader among them; they decide together where to go—a decision they make on the literal, honest-to-goodness fly. (Gonzalez, 2018)

Traffic Management

Multiple parties, both commercial and government, have begun to develop UAS traffic management (UTM) systems. Thales ECOsystem and AirMap propose communication with air navigation service providers or similar services. Single European Sky Air Traffic Management Research (SESAR) U-Space is taking a tiered approach, begin-

---

[9] See also Brown et al., 2016; Goodrich et al., 2013; Kolling, Sycara, et al., 2013; Kolling, Walker, et al., 2016; and Walker et al., 2014.

ning with registration and geofencing and moving on to communication and exchange of information with air navigation service providers. The National Aeronautics and Space Administration (NASA) UTM effort aims to support the development of "airspace design, corridors, dynamic geofencing, severe weather and wind avoidance, congestion management, terrain avoidance, route planning and re-routing, separation management, sequencing and spacing, and contingency management," effectively an air traffic control system for UASs (NASA, 2018).

These projects have the goal of incorporating UASs into existing or future air traffic management services. This capability has the potential to reduce restrictions on UAS use; they will no longer be obstacles to avoid but, rather, additional managed traffic in the sky. It also means that UASs will be easier to track and identify—including when monitoring for abnormal behavior—and nefarious actors will need to take on additional risk by using a registered UAS, take on a different risk by using an unregistered UAS, or have greater technological knowledge and the ability to hijack a third-party UAS.

Communication

SUASs communicate over multiple frequencies. Each platform may use different frequencies for radio control (transmitting and receiving) and its telemetry/video/sensor feed. Many platforms use common radio-control (RC) frequencies that have been allotted to hobbyist model aircraft flyers for decades, others use higher-frequency bands with spread spectrum became available more recently, and some even offer the capability to use satellite or cell signals.

The transmission ranges of these control methods may have a greater consequence for the overall range of the platform than battery-governed range limitations. However, there are several potential workarounds to these range limitations. A repeater platform may be used, such as another sUAS, that relays control information and content to and from the primary sUAS. This method introduces some lag in the signals and an added point of vulnerability in the form of the second sUAS, but these may be acceptable trade-offs for the extended range. And, in many areas, using a third-, fourth-, or fifth-generation (3G, 4G, or 5G) cell signal to control sUASs is a highly viable option, introducing some lag but greatly expanding the distance between user and the platform. Cell signal control also makes it very difficult to find the control signal for the sUAS because it blends in with normal cell signal traffic.

The Current State of the Art

UAS communication used to take place over a number of RC bands in the MHz region, commonly referred to as *narrowband control*. While common narrowband RC frequencies that have been used for decades still see some continued use with sUASs, the majority have moved to spread-spectrum bands: 2.4 GHz, 5.8 GHz, and other

GHz bands that offer significantly reduced interference and detectability.[10] Interference is expected and is compensated for; therefore, signal obfuscation is already included in these bands due to the use of Wi-Fi, wireless devices, and microwaves. Controllers using these bands eliminate common interference concerns that plague traditional narrowband control through the use of automatic channel allocation or spectrum hopping. Spread-spectrum control also has a higher bandwidth and increases the signal-to-noise ratio, both of which increase the link's data capacity, according to the Shannon-Hartley theorem.[11] Using these higher frequencies also results in a smaller antenna due to the shorter wavelengths, because RC antennas are sized proportionally to the wavelength of the frequency in use.

As Figure 2.8 shows that the 900–915 MHz band is the most common narrowband still in use, with 433 MHz also seeing some use (mainly by systems produced by European manufacturers).

But as Figure 2.9 shows, spread-spectrum use heavily outweighs narrowband use. This trend will continue as older systems using narrowband for control continue to become obsolete and new systems are added to the market.

Some UASs offer satellite communication (SATCOM) capabilities over the C-band, L-band, or S-band. Control may also take place over cell signal networks

Figure 2.8
Control and Telemetry Spectrum Use

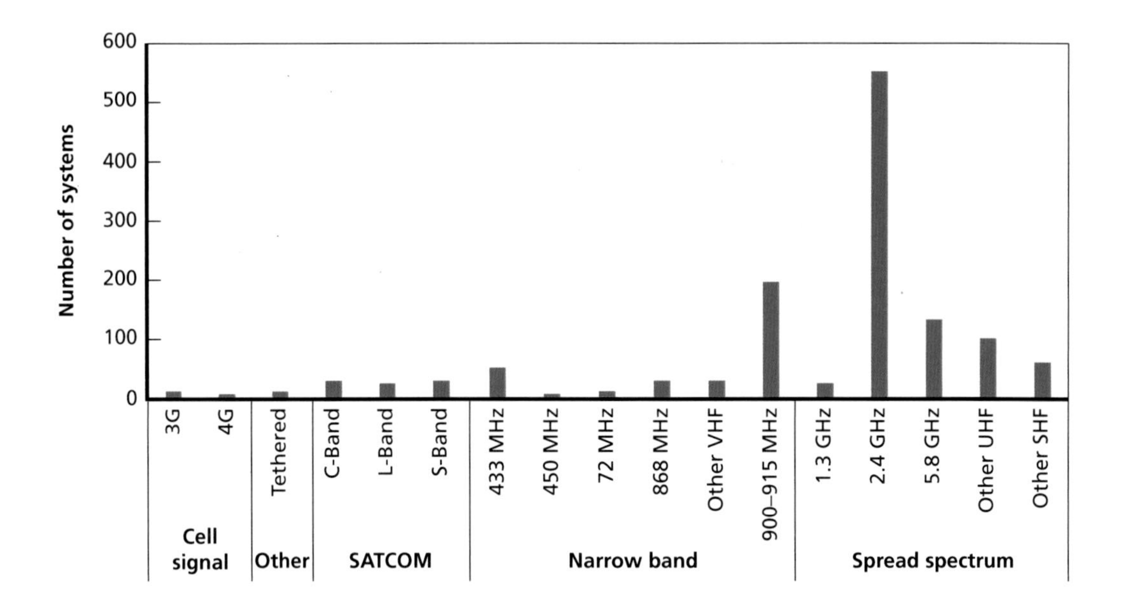

[10] The 2.4 and 5.8 GHz spectrum bands are used because they are industrial, scientific, and medical bands and have numerous uses.

[11] Channel capacity in bits per second = $\text{bandwidth} \times \log_2\left(1 + \frac{\text{signal}}{\text{noise}}\right)$.

Figure 2.9
Control and Telemetry General Spectrum Use

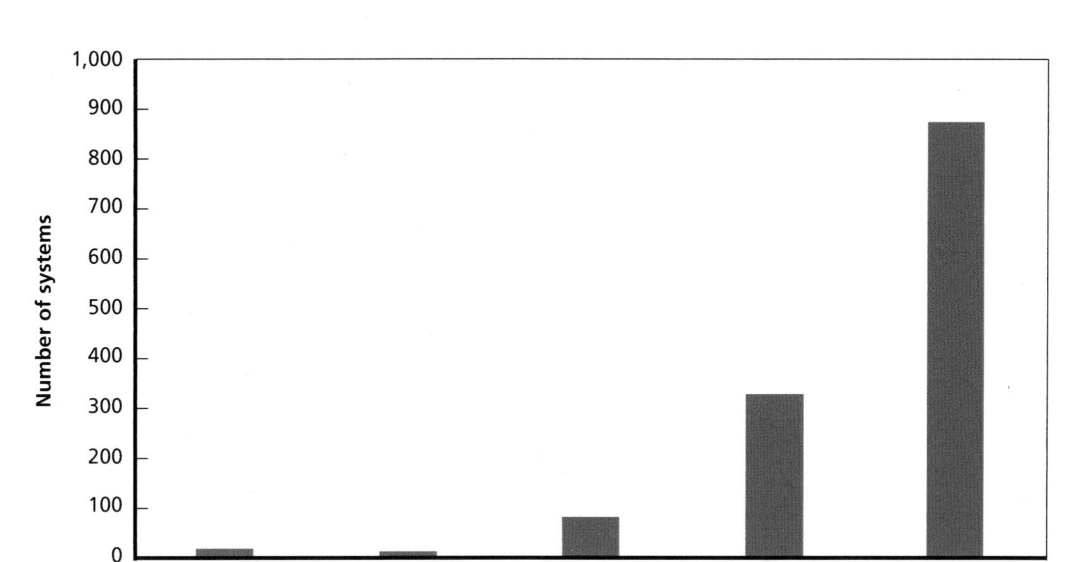

(3G and, more recently 4G, with potential 5G use when it is brought online). The latter of these is a potential boon for nefarious actors, as the ability to discern the control signal from ubiquitous normal traffic could prove to be particularly difficult in crowded urban environments.

Near-Future Limits

The 2.4 GHz band on which most RC systems rely is also used in a multitude of everyday items, from home Wi-Fi to cordless phones and microwave ovens. This crowding of the spectrum can (and has already been seen to) lead to non-RC interference issues, especially in dense urban environments. This crowding may force sUAS manufacturers toward less densely populated bands, such as 5.8 GHz and 1.3 GHz (which are currently used by sUAS but to a lesser degree than 2.4 GHz), or to other means of communication, such as cell signals. Another concern with spread-spectrum options is that their relatively high frequencies result in short wavelengths that are relatively more susceptible to absorption or multipath errors, again pushing manufacturers toward other control methods.

Although current SATCOM options may be of limited utility to nefarious users due to limited and expensive commercial satellite bandwidth, along with long latencies, this may change in the near future. Efforts such as OneWeb and SpaceX's Starlink are looking to implement large microsatellite constellations in low Earth orbit, making SATCOM a decidedly cheaper and faster option than it is today.

There are also trends in different bands for cellular control. Zeng, Lyu, and Zhang theorize,

> [F]or future 5G-and-beyond cellular systems, new technologies can be developed to address the unique UAV traffic requirement more efficiently. One possible solution is to use drastically different bands for uplink and downlink communications, such as the conventional sub-6 GHz for UAV downlink whereas the largely underutilized millimeter wave (mmWave) spectrum for UAV uplink. (Zeng, Lyu, and Zhang, 2018)

Beyond-Line-of-Sight Control

The ability to control an UAS via the phone network has been demonstrated. There are numerous online guides for how to build a drone that is controlled via the cellular network (Li, 2016). However, real-time control over current 4G networks has been seen to lack real-time responsiveness (Cheng, 2016). Variable traffic volume on phone networks can negatively affect remote control and can be random or a result of high-profile events (Mesbahi and Dahmouni, 2016). 5G networks are likely to mitigate this issue further. Also, autonomy can mitigate the need for low latency over cellular networks. One example is the PX4 open-software autopilot flight stack for managing the open-hardware project Pixhawk, which enables autopilot (PX4, 2018).

PX4 developers describe the capabilities of the autonomous flight modes as follows:

> Flight modes provide different types/levels of vehicle automation and autopilot assistance to the user (pilot). *Autonomous modes* are fully controlled by the autopilot, and require no pilot/remote control input. These are used, for example, to automate common tasks like takeoff, returning to the home position, and landing. Other autonomous modes execute pre-programmed missions, follow a GNSS beacon, or accept commands from an offboard computer or ground station. (PX4, 2018)

PX4 has been integrated into companion autopilot hardware, such as Pixhawk, which provides "readily-available, low-cost, and high-end, *autopilot hardware designs* to the academic, hobby and industrial communities" (PX4, 2018). There are instructions for integrating Pixhawk with an Iridium/RockBlock satellite communications for about $13 a month and a latency of approximately ten to 15 seconds. Furthermore, there are real-time kinematic GNSSs to provide centimeter-level position accuracy. Future trends include miniaturization of the systems and increases in performance (Pixhawk, undated).

Radio Navigation Denial

Satellite-based radio navigation systems (commonly GPS but also including the Russian system GLONASS, the Chinese BeiDou, the European Union's Galileo, and others) are how nearly all UASs keep track of their positions. However, UASs also have fail-safe options and backups that complement radio navigation inputs and provide an alternative method of positioning when radio navigation is not available. Traditionally, radio navigation–denied position finding has relied on other onboard sensors and information to cover the lack of true position knowledge provided by radio navigation systems. SUASs contain inertial navigation sensors for control purposes, and some can repurpose these sensors for navigation during periods of denied coverage, though as with all inertial sensors, they are susceptible to integration drift over time.[12] More complex navigation systems rely on creating or referencing a known map or characteristics of the area that the UAS is traversing, such as terrain following, simultaneous localization and mapping, or target following, and are also sometimes used by sUASs, often as a complement to radio navigation.

Other, novel ways of tracking position can be supposed. Manned aircraft sometimes use very high frequency (VHF) omnidirectional range (VOR) navigation systems. These systems use terrestrial beacons broadcasting in the 108–117.95 MHz range to fix their position, in much the same way one would use the known locations of satellites in a GNSS. Such a system could be implemented on an sUAS at the cost of more equipment and power draw. Although previous discussion on SDR use suggests that this may be a minimal concern if the SDR can simply ignore this function until required. Manned aircraft also use ADS-B to report their positions via satellite navigation and periodically broadcast it. ADS-B is intended for use by other aircraft and air traffic controllers to facilitate safe travel, but if a sufficient number of aircraft were within transmittal range, a UAS could use these broadcasts to fix its location. This is a more difficult problem than that of GNSS and VOR due to the biased clocks of the aircraft, but clever use of known conditions, such as aircraft type or geographic closeness, could alleviate this problem to give a sufficiently accurate position.

FOAM

FOAM is one novel approach that has been proposed to enable operations in environments where radio navigation is not possible. FOAM is an open, decentralized protocol for building a consensus-driven map of the world using blockchain to verify inputs. It is still in development but may be able to provide basic navigation and timing capabilities in a localized area (FOAM, undated).

A blockchain in its simplest form is a digitized ledger on which entries are linked using a cryptographic process. Blockchains require no central authority and can be

---

[12] Small errors in the sensors' measurement of acceleration and angular velocity are compounded in the calculations for platform speed and position. Because each subsequent measure of position uses the previously measured position value in its calculation, these errors compound continually over time until the measure can be reset.

maintained over peer-to-peer networks with intermittent connectivity by each individual user.

FOAM is an attempt to apply blockchain technologies to enable location-dependent smart contracts on the Etherium network through a peer-to-peer location and mapping system. According to recent white papers (FOAM, 2018a, 2018b), FOAM intends to set up a network of "zones" that are monitored by "zone authorities," "zone anchors," and "validators." Zone authorities maintain a precision digital clock system, a precisely known location, connection to the network's blockchain, and a capability to transmit and receive RF messages, with four zone authorities required to designate a zone and maintain synchronized timing. Zone anchors are similar to zone authorities but receive all timing information from zone authorities and do not support synchronization. Validators are agents on a network that verify output from zone anchors and zone authorities.

One other agent on the network is a "mobile beacon" that is not in an established location. When within a zone, a mobile beacon makes a request to authorities and anchors in a zone using a "presence claim" requesting its location and offering a fee for verification (for example, "I am 123 and will pay $5 if you tell me where I am right now"). Anchors and authorities that agree to accept the fee will send a message to the mobile beacon (similar to GNSS) and request a message back (unlike GNSS). The time of arrival is used to determine the difference between the mobile beacon and each anchor and authority. A minimum of four zone anchors or zone authorities are required to establish a position in 3D space, but more can be requested for higher fidelity. Validators then use the distances to evaluate where "123" is and assess whether any other entity is malfunctioning or intentionally giving bad data. Once this process is complete, the location of 123 is reported to the blockchain, any contracts that require the location are executed, and every properly functioning and honest participant is paid a fraction of the fee.

This system has several advantages over existing positioning technology. It can provide precise timing and positioning in a local area due to the synchronized zone. The only required external interface is with the global Etherium blockchain over the internet (FOAM, 2018a, 2018b). The current FOAM system is designed to be transmitter-agnostic; however, the current LoRaWAN protocol has been chosen because it is in an unlicensed low-frequency spectrum with a five- to 15-kilometer range and low power requirements (The Things Network, undated; LoRa Alliance, undated; Semtech, undated). Existing Link-16 antennas could work on the LoRaWAN frequency, possibly requiring only software changes to enable capabilities on existing platforms with Link-16 because they have similar frequency ranges (see Wikipedia, 2018b). The FOAM protocol could provide a degree of anonymity because the actual identity of the mobile beacon is not broadcast, only the presence claim.

However, this system has significant shortcomings. The first is the chosen RF protocol. LoRaWAN is not designed for high data rates and takes several minutes to transmit messages at the highest speeds. Additionally, because the FOAM protocol requires several steps to update (broadcast presence claim, zone anchor response, mobile beacon response, validation, post to the Etherium blockchain), that could also take several minutes. While this is perfectly fine for locating stationary and slow-moving objects, it would be difficult to determine the position of fast-moving objects with the current FOAM protocol. Finally, FOAM only provides anonymity, not privacy. Anyone on the network would see that someone has made a presence claim, just not who. If there is a limited number of a particular type of agent in an area, such as low-flying UASs, identifying that the mobile beacon is a U.S. UAS would be trivial.

Zones are vulnerable to several types of faults. If only four zone authorities are present, loss of a single authority—for reasons unintentional (e.g., power outage) or intentional (e.g., false message)—would result in a loss of synchronization for the entire system. Additionally, only one in three validators needs to be wrong to prevent the validation of presence claims. While none of this would be a problem in a robust network with many zone authorities, zone anchors, and validators, other blockchain-enabled networks, such as Bitcoin, have become highly centralized with a few market participants because economies of scale favor those who can purchase hardware and energy at wholesale prices (Homakov, 2017). In these instances, a single participant could exploit, disable, or eliminate trust in the network by altering the timing of messages or misstating actual positions. A mobile beacon could also spoof its position if it were to alter timestamps sent to each zone anchor and zone authority in a coordinated manner, or else the validators would identify the fault similarly to how validators would identify faulty zone authorities and anchors.

Although FOAM as a service may not be optimal as a GNSS alternative for UASs, some of its underlying technology could be. The overall RF location method is more resistant to jamming and disruption than GNSS, and, if paired with a better RF protocol, it could provide highly accurate locations and timing. If paired with prepositioned or deployable mobile transmitters—ignoring blockchain portions of the FOAM system—this could provide local precision navigation and timing in a GNSS-degraded environment.

Overall, the FOAM protocol shows promise in providing peer-to-peer proof of location to enable smart contracts via blockchains, such as Etherium. It provides a way of anonymously providing secure timing and location services for stationary and slow-moving objects without relying on GNSS. However, FOAM has some inherent difficulties providing real-time location services for fast-moving objects, and it has some significant security vulnerabilities. Individual components of FOAM have the potential to provide GNSS-like position, navigation, and timing capabilities for mobile systems, such as UASs, but the FOAM protocol itself is probably not an effective substitute for GNSS capability.

Propulsion System and Power Supply

The propulsion system, along with the sUAS's power supply, dictates operational range and endurance. We developed equations to estimate the endurance and range of rotary-wing a UAS. Testing against several commercial rotorcraft, we found these equations to be reasonably accurate. At the end of this section, we discuss the various current and future power supplies that may be used by an sUAS.

Endurance Estimation

Endurance is defined as the total amount of time an aircraft can stay in flight. For larger fixed-wing aircraft, which usually come fitted with fuel-based propulsion systems, endurance depends on the amount of fuel a craft can carry. However, electric rotary-wing UASs do not carry fuel. Their propulsion systems run off of battery power. Keeping this in mind, we developed an equation for sUAS endurance by drawing parallels between aircraft endurance and battery life. The end result was an equation that incorporated characteristics from the sUAS's propulsion and electrical systems:[13]

$$\text{Endurance} = \frac{B_C}{I_H}\eta, \tag{2.3}$$

where B_C = battery capacity, I_H = current draw at hover, and η = motor efficiency.

The equation is based on a set of assumptions. First, we assumed that maximum endurance is realized at hover.[14] We also assumed that the amount of lift required to achieve hover is equal to the mass of the aircraft. Finally, we assumed 75-percent motor efficiency.

We tested this endurance equation against several versions of the DJI Phantom. We were able to accurately predict the maximum endurance of each sUAS within a small margin of error. The results are shown in Table 2.2.

It is important to note that we approximated the amount of current required for hover using the maximum lift and maximum continuous current allowed for each craft. This approach could be the reason for the small margin of error seen in Figure 2.10.

[13] Battery capacity and current assume electric propulsion. A similar measure should be inserted for other power sources.

[14] While some small forward speed would result in greater endurance than the aircraft at hover, this effect is likely to be negligible due to the generally low speeds of sUASs. Comparing our estimates with reported values shows this to be true.

Table 2.2
Factors Used in Calculating Battery Life

UAS Model	Motor Model	Number of Motors per Craft	UAS Mass (g)	Thrust Required per Motor for Hover (N)[a]	Max Thrust per Motor (N)	Max Continuous Current (A)	Current Required for Hover (A)	UAS Battery Capacity (Ah)	Estimated Efficiency	Estimated Endurance (minutes)
Phantom 1	E300	4	1,000	2.5	5.9	15	6.3	2.2	0.75	15.8
Phantom 2	E300	4	1,300	3.2	5.9	15	8.1	5.2	0.75	28.8
Phantom 2 Vision	E300	4	1,160	2.8	5.9	15	7.3	5.2	0.75	32.3
Phantom 3	E305	4	1,216	3.0	8.3	20	7.2	4.48	0.75	28.2
Phantom 3 SE	E305	4	1,236	3.0	8.3	20	7.3	4.48	0.75	27.7
Phantom 4	E305	4	1,380	3.4	8.3	20	8.1	5.35	0.75	29.7
Phantom 4 PROV2.0	E305	4	1,375	3.4	8.3	20	8.1	5.87	0.75	32.7

[a] DJI gives thrust values in units of mass (kilograms). We converted to a more traditional use of force (newtons) by multiplying by gravitational acceleration.

Figure 2.10
Endurance Estimation Algorithm Tested Against Reported DJI Phantom Specifications

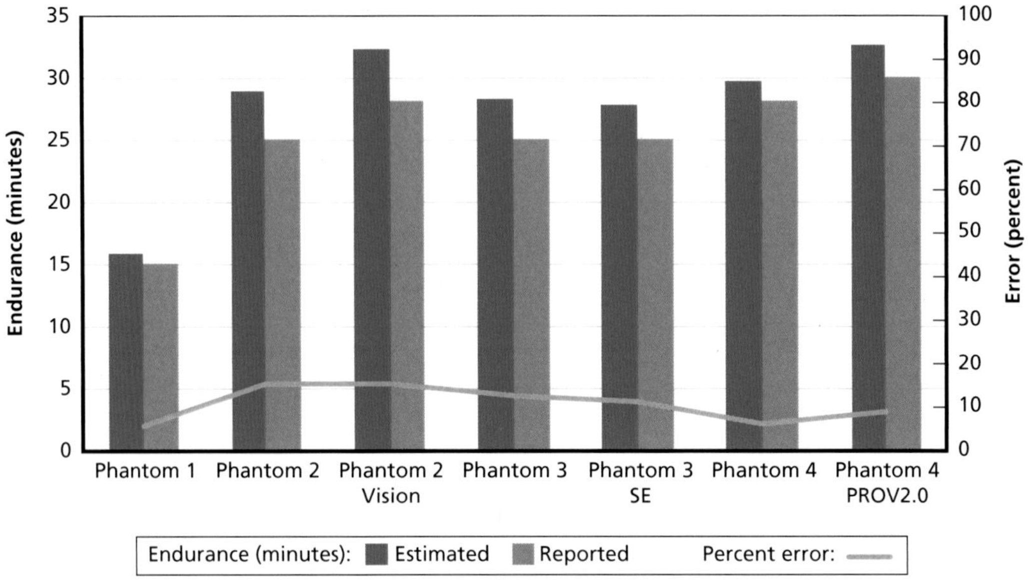

Endurance (minutes)
35 30 25 20 15 10 5 0
Error (percent)
100 90 80 70 60 50 40 30 20 10 0
Phantom 1 Phantom 2 Phantom 2 Vision Phantom 3 Phantom 3 SE Phantom 4 Phantom 4 PROV2.0
Endurance (minutes): Estimated Reported Percent error:

Range Estimation

Range is defined as the maximum ground distance an aircraft can cover in flight. Measuring range for a UAS requires two pieces of information: the UAS's velocity and the endurance of the craft at that specific velocity, because velocity and endurance are not independent in the case of a UAS. To achieve a certain velocity, the craft's motors must be operating at or above a certain level. A higher level of operation requires a higher amount of current, and the increased current draw decreases the aircraft's endurance. Therefore, to estimate sUAS range, we must first determine the mathematical relationship between velocity and endurance. As it turns out, the two factors can be linked through the force of lift. Equation 2.4 outlines how velocity can be calculated using lift, and Equation 2.5 does the same for endurance.[15]

$$V = +\sqrt{\frac{L\cos\theta}{\rho C_d A}}, \tag{2.4}$$

[15] Equation 2.4 is derived from the standard drag equation and assumes that drag is equal to lift. Battery capacity and current in Equation 2.5 assume electric propulsion. A similar measure should be inserted for other power sources. Note that the effective surface area of the UAS is a flat plate whose size would result in equivalent drag between the flat plate and UAS under investigation.

where V = velocity, L = lift force, θ = aircraft angle of attack, ρ = air density, C_d = drag coefficient, and A = effective surface area.

$$E = \frac{B_C T}{IL\cos\theta}\eta, \tag{2.5}$$

where E = endurance, B_c = battery capacity, T = thrust, I = current, L = lift, θ = aircraft angle of attack, and η = engine efficiency.

Equations 2.4 and 2.5 can be combined to give us an equation for the UAS range:

$$\text{Range} = \text{endurance} \times V = \frac{B_C T}{IL\cos\theta} \times \sqrt{\frac{L\cos\theta}{\rho C_d A}}. \tag{2.6}$$

Rotary-wing sUASs, it should be noted, have higher drag and a larger effective surface area than fixed-wing aircraft in forward flight. This leads to an intuitive result: Fixed-wing aircraft are often faster, with greater range and endurance, than their rotary-wing counterparts. However, rotorcraft offer the advantages of VTOL, hover, and low/slow flight in exchange for reduced efficiency in forward flight. Using Equation 2.6, we estimated maximum range for several generations of the DJI Phantom (standard series only). Our results, which can be seen in Table 2.3 and Figure 2.11, were found to be reasonably accurate.

It is important to note that estimating maximum range required some assumptions. We assumed that air density was measured at sea level. Additionally, we approximated the coefficient of drag and effective surface area of each UAS using each craft's measured dimensions (as reported by the manufacturer). Because of the quadrotor design of this sUAS platform, we assumed that the Phantoms suffer higher drag coefficients and effective surface areas than their fixed-wing counterparts. Our estimates in Figure 2.11 approximately match the reported range values of the tested DJI models.

Table 2.3
Estimated Range Based on Endurance

Model	Max Speed (mph)	Estimated Endurance (minutes)	Estimated Range (miles)	Reported Range (miles)	% Error
Phantom 1	22	16	1.8	2.5	28
Phantom 2	33	29	4.3	4.5	4
Phantom 3	35	28	5.2	6.0	13
Phantom 4	45	30	7.9	8.0	1

Figure 2.11
Range Estimation for DJI Phantom Series (Standard Only)

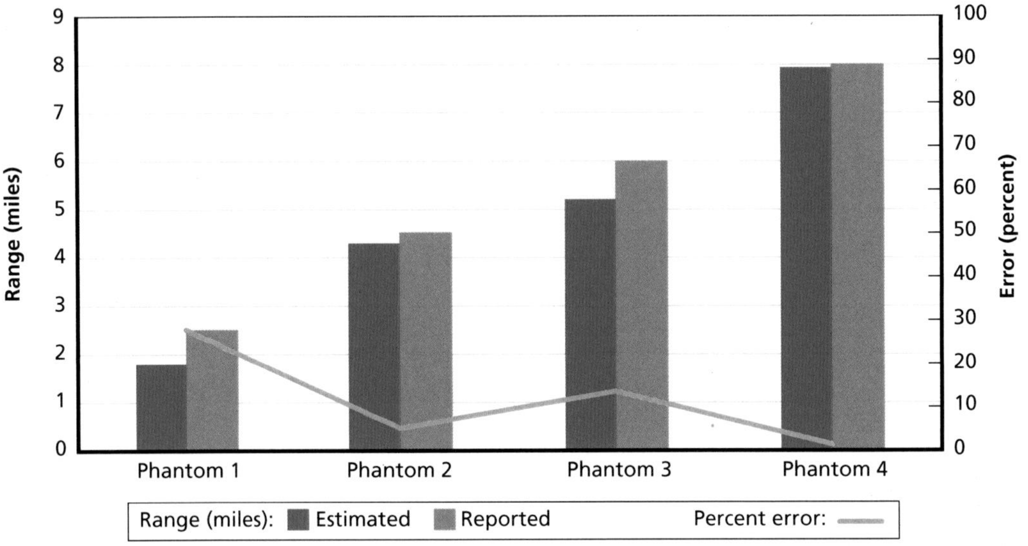

Power Supply Technology
Current State

The majority of sUASs in use today, in terms of number of models and number of in-use aircraft, are battery-powered. Batteries offer simple, inexpensive energy production for an sUAS. They also reduce noise and offer easier integration between the power source and onboard electronics. Internal combustion–powered sUASs do have some specific uses, especially for missions that are not concerned with detectability and that require only speed or endurance (often featuring fixed-wing UASs).[16] However, they are a decided minority; fewer than 25 percent of the models in our database were powered by internal combustion piston or turbine engines. Even more rare were hybrid engines that can switch or supplement one type of propulsion with another. These engines enable relatively high ranges and speeds, as well as the ability to run quietly on electric power, in much the same way hybrid automobiles operate. The obvious cost is in the increased complexity and weight of a hybrid engine.

Battery power may come in many forms. Lithium-polymer (LiPo) batteries offer relatively high energy density compared to other common battery types such as Li-ion, nickel cadmium, or nickel metal hydride. As such, more than 90 percent of battery-powered models in our database use these batteries.

[16] Internal combustion engines may see even greater advantages if oxygen-carrying fuels are implemented, but, again, concerns about noise, mechanical complexity, and detectability would persist.

Batteries are traditionally recharged via ground power when the aircraft lands, but solar and tethered systems are alternative options. Solar power can recharge batteries midflight, but it tends to be relatively slow and inefficient, and it is mainly used on long-endurance gliders. Tethered electric power allows long-endurance stationkeeping by being constantly plugged into a source of power, but the UAS is unable to move beyond the tether's range. These both provide limited utility to nefarious users who will typically seek out targets where they are not able to set up ground power and require more stealth and speed than a solar glider can provide.

Near-Future Advancements

Like all other aircraft manufacturers, sUAS manufacturers are always seeking ways to alleviate size, weight, and power concerns and increase range and endurance. This could be achieved to some extent with improvements to battery technology, as was the case with the development of lithium-based batteries or the introduction of newer battery types, such as solid-state and flow batteries.

Lithium-Based Batteries

More than 90 percent of battery-powered sUAS models being manufactured today use LiPo batteries. An overwhelming amount of research has focused on Li-ion batteries; Li-ion batteries are almost identical to LiPo batteries and will likely replace the latter as an sUAS power source in the near future.[17] Li-ion batteries can be found in almost every modern electronic device, a fact that can be attributed to the battery's high capacity and rechargeability. However, this type of battery has several limitations, including an inability to operate effectively at extreme temperatures, degradation with age, and, perhaps most importantly, safety concerns regarding the encased electrolyte component. These factors mitigate the Li-ion battery's power output, pushing manufacturers to explore different options in an attempt to improve battery technology. One potential solution involves new electrode materials. Sila Nanotechnologies, a California-based battery startup, has been researching the incorporation of silicon-based nanoparticles into Li-ion anodes (Temple, 2018; Goode, 2018). This new type of anode can store a larger number of lithium ions than the traditional graphite-based anode, allowing for a reported 20-percent increase in battery energy density (Morgasinski et al., 2010). Enovix and Enevate, two other California-based startups, are reportedly looking to improve battery life by creating silicon-based anodes (Temple, 2018).

Rather than simply modifying the current electrode materials in Li-ion batteries, some researchers are experimenting with entirely new materials, such as sulfur or

---

[17] Of the 545 currently manufactured battery-powered models in our database for which we have battery-type information, 514 exclusively use LiPo batteries.

LiPo batteries use slightly modified versions of the Li-ion battery's separator and electrolyte; this allows the LiPo battery to have a flatter, more rectangular shape, but its specific capacity and applications are essentially identical to those of a Li-ion battery.

lithium.[18] These batteries have been dubbed lithium-metal (Li-metal) batteries. The electrodes of Li-metal batteries are composed of pure metals, as opposed to the Li-ion battery's mixed compounds. The most promising Li-metal candidate is the lithium-sulfur (Li-S) battery. This battery type is still in the experimental stage, and its drawbacks include a very poor life cycle. The upsides of the Li-S battery, however, include a high capacity and a low manufacturing cost (Sims and Crase, 2017). The effectiveness of Li-S batteries has already been demonstrated in the Airbus Zephyr, a fixed-wing UAS that was initially developed by QinetiQ. In March 2010, the Zephyr shattered the world record for the longest unmanned flight, flying for more than 336 hours. The aircraft was powered by solar energy during the day and used Li-S batteries to operate at night (Sion Power Corporation, 2010). This flight demonstrates the potential of Li-S battery use in UASs, but, in this case, the batteries were customized specifically for the Zephyr and its high-altitude flight. Commercial use of Li-S batteries could still be five to ten years away, and it is unclear whether the new chemistries are safe for commercial use (Choi and Aurbach, 2016).

Another alternative to the Li-ion battery is the lithium-air (Li-air) battery, sometimes known as the lithium-oxygen or metal-air battery. This battery type creates a reaction between lithium and oxygen particles, resulting in an extremely high-energy-density system. Theoretically, the energy density of the Li-air battery is 100 times greater than that of the Li-ion battery, putting Li-air batteries on par with gasoline (Buchmann, 2018). Although they were first demonstrated in the early 1990s, Li-air batteries are still an immature technology and have several problems that make them unsuitable for sUAS use. For example, the batteries have an extremely short life cycle, and, for now, several iterations of the battery have proved to be nonrechargeable. However, recent research has shown that the next generation of Li-ion batteries may overcome these hinderances. Chemists at the University of Waterloo in Canada have created a Li-air battery that can be recharged up to 150 times; this invention is still a long way from commercialization, however, as the battery must be heated to 150 degrees Celsius before operation (Xia, Kwok, and Nazar, 2018).

The standard Li-ion battery contains a nonaqueous liquid or nonaqueous gel electrolyte. These mixtures are composed of lithium-based salts dissolved in organic carbonates. As mentioned earlier, this electrolyte substance comes with certain safety concerns (Brunning, 2016). The current technology also features some inefficiencies, as reported by researchers from SLAC National Accelerator Laboratory and Lawrence Berkeley National Laboratory. Their recently published findings explore how lithium ions travel through the battery material and showcase the inefficiencies of the current pathways (SLAC National Accelerator Laboratory, 2018). Researchers are studying several solutions to these problems, the most promising of which may be the development

[18] Current Li-ion batteries do not use lithium as an electrode but, rather, store lithium ions in a porous carbon-based material, such as graphite (Mekonnen, Sundararajan, and Sarwat, 2016).

of new battery electrolytes. Aqueous electrolytes, which are water-based instead of carbonate-based, have been shown to improve Li-ion battery performance and alleviate the previous safety concerns (Suo et al., 2015). Ionic liquid electrolytes have garnered a similar reputation, but their viscous nature could reduce battery conductivity (Li, 2016). Researchers are also looking into solid electrolyte alternatives, as we discuss below. Ultimately, the battery's electrolyte does little to increase battery capacity.

Flow Batteries

Flow batteries are an alternative for sUASs, but for the most part, are currently only used for electrical grid applications. Flow batteries seek to replace the components of a traditional battery with purely liquid components. In place of the anode and the cathode, this type of battery features the anolyte and catholyte, liquid slurries with dissolved chemical components. As the battery is used, these slurries are depleted. Fortunately, flow batteries offer the benefit of near-instantaneous refueling by simply refilling their tanks. On the other hand, however, flow batteries also require mechanically complex pumping structures to move the slurries. Flow batteries are currently being developed for electric cars and could lead to a 50-percent increase in energy density over current technology (NanoFlowcell, undated). Additionally, NASA is investigating nanoelectrofuel flow batteries for aircraft (Warwick, 2018), so sUASs may soon be equipped with miniaturized flow batteries. It remains to be seen if any of these solutions are economically viable.

Solid-State Batteries

As mentioned earlier, researchers are looking into the possibility of replacing the current Li-ion battery's liquid electrolyte with a solid electrolyte. Doing so would remove all liquid components from the battery, creating what is called a *solid-state battery*. Researchers have explored the use of ceramic- and glass-based substances for this purpose, but Massachusetts-based Ionic Materials has opted for another option: a conductive plastic polymer. The company's research shows that using this polymer, which is a flame-retardant material, would improve battery safety in addition to increasing battery life (Jülich Forschungszentrum, 2018). If this innovation is coupled with improvements to Li-S batteries, sUAS power sources could see a large spike in battery capacity. Solid-state batteries are not expected to be commercially available soon but are expected to at least double the energy density of traditional batteries (Goode, 2018; Kato et al., 2016).

Fuel Cells

Some sUASs have begun using fuel-cell batteries in place of lithium-based batteries—and to good effect. In our database, sUASs with fuel cells demonstrated greater endurance capabilities. Fuel cells are an older concept, having been invented in the 1800s. Compared with Li-ion batteries, fuel-cell batteries offer a much higher energy density. Unfortunately, fuel cells, which can be either hydrogen-based or methanol-based, suffer several complications that have prevented their use in modern aircraft. Hydrogen-based fuel cells require a special container that is expensive, heavy, complex, and prone to breakage (Urbonavicius et al., 2017). Methanol fuel cells have been shown to resolve some of these concerns; they do, however, feature other problems related to design, fabrication, and operation that are unresolved (Basile and Dalena, 2017).

Laser Power

In laser-powered UASs, energy is supplied from offboard sources via lasers directed. This enables electric operation and potentially much longer-range operation, provided an energy source is available along the way. However, it is unlikely that this technology will be leveraged in commercial sUASs within the next few years (Adams, 2018).

Exploration of Future sUAS Performance

The battery advancements discussed here are expected to occur in distinct stages over the course of several years. Using this information, we can establish an estimated timeline for each advancement and incorporate the research of battery technology into our current endurance and range models using the DJI surrogates. It is possible that flow and Li-metal batteries will reach the point of technological readiness suitable for sUASs; for the purposes of this exploration, we assumed five years from the time of this research in 2018. A similar jump could be seen in the year 2028, when researchers may be able to scale and reproduce effective solid-state batteries. Based on earlier models, both the endurance and range of sUASs could double in that time, as shown in Tables 2.4 and 2.5. One battery type that we did not account for was the Li-air battery, which, as mentioned earlier, could improve battery capacity by a factor of 100. Because of the immaturity of this technology, it is difficult to predict when the Li-air battery will become commercially available, and it is highly unlikely that this will occur within ten years.

Some sUASs are already showing significant improvements. Impossible Aerospace has introduced a novel concept wherein the aircraft structure of its multirotor sUAS

Table 2.4
Possible sUAS Endurance Increases Over a Ten-Year Period

Model	Traditional Li-ion Battery (2018)	Li-Metal and Flow Batteries (expected ~2023)	Solid-State Battery (expected ~2028)
Phantom 1	15.8	19.0	31.7
Phantom 2	28.8	34.6	57.6
Phantom 3	28.8	33.8	56.4
Phantom 4	29.7	35.6	59.3

NOTE: All values in minutes.

Table 2.5
Possible sUAS Range Increases Over a Ten-Year Period

Model	Traditional Li-Ion Battery (2018)	Flow Battery (expected ~2023)	Solid-State Battery (expected ~2028)
Phantom 1	1.8	2.2	3.6
Phantom 2	4.9	5.1	8.5
Phantom 3	5.2	6.2	10.4
Phantom 4	7.9	9.5	15.8

NOTE: All values in miles.

is filled with battery cells, allowing for a large battery capacity that doubles as a structural support. While it is possible that its aircraft falls within the scope of our models, the Impossible Aerospace UAS claims an endurance of two hours with this innovative design (Impossible Aerospace, undated). Horizon Energy Systems has taken a similar approach of using the airframe to store fuel, but instead of batteries, it holds hydrogen fuel cells (Unmanned Systems Technology, undated).

There are some researchers who, instead of improving the sUAS battery technology, have focused their sights on improving small-engine technology. These researchers have been exploring methods to shrink down internal combustion engines to incorporate them into quadcopter UASs. Solely reducing the size of the engine creates several problems for the internal mechanics, however (Griebel, 2010). At this time, it is not possible to determine the impact that this would have on sUAS performance (Sher and Sher, 2011).

Software Security

As of 2016, there were more than two dozen documented instances of UAS exploitation (Walters, 2016). This is in addition to academic demonstrations.

Command, Control, and Communications Security

Communications and nodes for transmitting information inherently increase the vulnerability of detection as well as avenues through which one can track, send information to deauthenticate original controller leading to a takeover, and send deletion or new "go to" commands for final attack. Those communications can be detected, potentially listened to for location and intention information, then seized to cut off sUAS from higher command, and used to send misinformation to end mission or redirect sUAS to a known location. This whole process is often referred to as *dronejacking*. Johns Hopkins University exploited the vulnerabilities of a hobby drone to show how simple the process can be:

> An "exploit," explained Michael Hooper, one of the student researchers, "is a piece of software typically directed at a computer program or device to take advantage of a programming error or flaw in that device." In the team's first successful exploit, the students bombarded a drone with about 1,000 wireless connection requests in rapid succession, each asking for control of the airborne device. This digital deluge overloaded the aircraft's central processing unit, causing it to shut down. That sent the drone into what the team referred to as "an uncontrolled landing."
>
> In the second successful hack, the team sent the drone an exceptionally large data packet, exceeding the capacity of a buffer in the aircraft's flight application. Again, this caused the drone to crash.
>
> For the third exploit, the researchers repeatedly sent a fake digital packet from their laptop to the drone's controller, telling it that the packet's sender was the drone itself. Eventually, the researchers said, the drone's controller started to "believe" that the packet sender was indeed the aircraft itself. It severed its own contact with the drone, which eventually led to the drone making an emergency landing.
>
> "We found three points that were actually vulnerable, and they were vulnerable in a way that we could actually build exploits for," [Lanier A.] Watkins [who supervised the research] said. "We demonstrated here that not only could someone remotely force the drone to land, but they could also remotely crash it in their yard and just take it." (Sneiderman, 2016)

Source Code Security

DJI suffered for years from developers who placed DJI source code, which included keys for decrypting the flight control software, in a public GitHub location that had been forked, or duplicated and modified, elsewhere (Corfield, 2018). The Phantom 2 was riddled with software insecurities that were easily leveraged by even novice hackers to take control of the aircraft (Trujano et al., 2016). According to Trujano et al. and the Massachusetts Institute of Technology, "The Phantom 3 Standard is more secure than its predecessors, yet still fails to meet the unique demands of airborne technology in the context of security" (Trujano et al., 2016). Dey et al., concluded an analysis of the Phantom 4 Pro with the following:

> The P4P is one of the most secure and robust drones available in the commercial market. Although DJI attempted to develop the P4P, so that it is less vulnerable than its predecessors, it still needs further work and as well as comprehensive security analysis. (Dey et al., 2017)

One conclusion that can be drawn through these studies is that DJI continues to increase its security and raise the bar for the level of resources required to disrupt, deny, or disable sUASs.

Summary of Findings from the Technology Capability Assessment

We developed a database of sUAS performance from a variety of sources, including AUVSI database of air platforms, *Jane's Defence: Air Platforms*, manufacturer-provided data, and academic papers. In the process, we made the following observations:

- Most sUAS platforms today are either fixed-wing or rotary-wing.
- Carbon fiber composite is the most popular airframe material by a factor of six.
- Miniature radar payloads have been developed to work on sUASs as small as 1.65 lbs with a range of 3 km at 30 W.
- Miniaturized hyperspectral and LIDAR sensors are being developed but appear less mature than radar for sUAS applications.
- GNSS jamming and spoofing will continue to be attractive applications for nefarious sUAS users due to the limited power required to jam or spoof and the ability of an sUAS to get into unobstructed positions to maximize its impact.
- Mobile phone jamming will likely remain outside the capability of nefarious sUAS users due to the power requirements necessary to be effective.
- The Walden curve shows a tenfold increase in processing capability of A/D converters, enabling sUASs to collect more electronic signals.
- Increasing computing power and parallelism from FPGAs and GPUs is enabling the development of beam-forming antenna technology at a fraction of the size, weight, and power historically required, enabling sUASs to perform surveillance missions that were once suitable only for larger aircraft.
- Drone swarms commanded partly by automation and partly by simple "flocking" rules are more likely than fully autonomous swarms in the near future.
- Advances continue to be made in enabling human operators to control multiple sUASs concurrently.
- Spread-spectrum use of communication frequency continues to be favored over narrowband by a factor of two.
- There are many how-to guides on the internet describing how to fit an sUAS with a cellular network transceiver for remote control. Users likely experience latency on today's 4G network, but advances in this area will make it more difficult to identify sUASs via communication links.
- Open-source software to manage automated route flying, autonomy, integration with GNSS for precise location, and numerous other features is available today.
- Battery technology is notorious for overpromising, but we estimate that flow batteries could improve the range of the DJI Phantom 4, for example, by 20 percent, and solid-state batteries by 100 percent.
- Awareness of software security in sUAS function is increasing, but there have been numerous examples of software insecurity (e.g., loss of source code control, lack of encryption), enabling drone-jacking and suggesting that these problems will continue, at least in the near term.

CHAPTER THREE

Performance Analysis

In this chapter, we explore differences in sUAS performance across five metrics: maximum range, endurance, payload weight, speed, and maximum altitude.

Data Completeness

The AUVSI database did not have complete information for any of these fields. After we filled in data that we found by consulting available documentation, the endurance field had the lowest percentage of missing data (less than 10 percent of the data were missing), and the maximum altitude field had the most missing data (nearly half of the data were missing). There were multiple fields for speed, including speed, maximum speed, and cruising speed. We created a combined speed variable that consisted of maximum speed, speed (if maximum speed was missing), or cruising speed (if maximum speed and speed were both missing). The frequencies of missing performance data are shown in Figures 3.1 and 3.2. Given the extremely high rates of missing data for maximum ascent and descent speeds, we did not include those metrics in our analysis; however, they may be worth exploring in the future.

We combined data on lift generation (fixed-, rotary-, and hybrid-wing) and propulsion types (internal combustion, battery electric, solar electric, tethered electric, hybrid, and jet engine) to generate classes of platforms (see Figure 3.3). "Other" types of platforms included lighter-than-air models, gliders, ornithopters, insectothopters, and flapping-wing models. The vast majority of platforms in the database are rotary-wing/battery electric (41 percent), fixed-wing/battery electric (35 percent), or fixed-wing/internal combustion (12 percent).

Maximum Range and Endurance

Figure 3.4 shows the reported ranges of sUASs by their joint airframe and propulsion type.

Figure 3.1
Missing Performance Metric Data in the Database

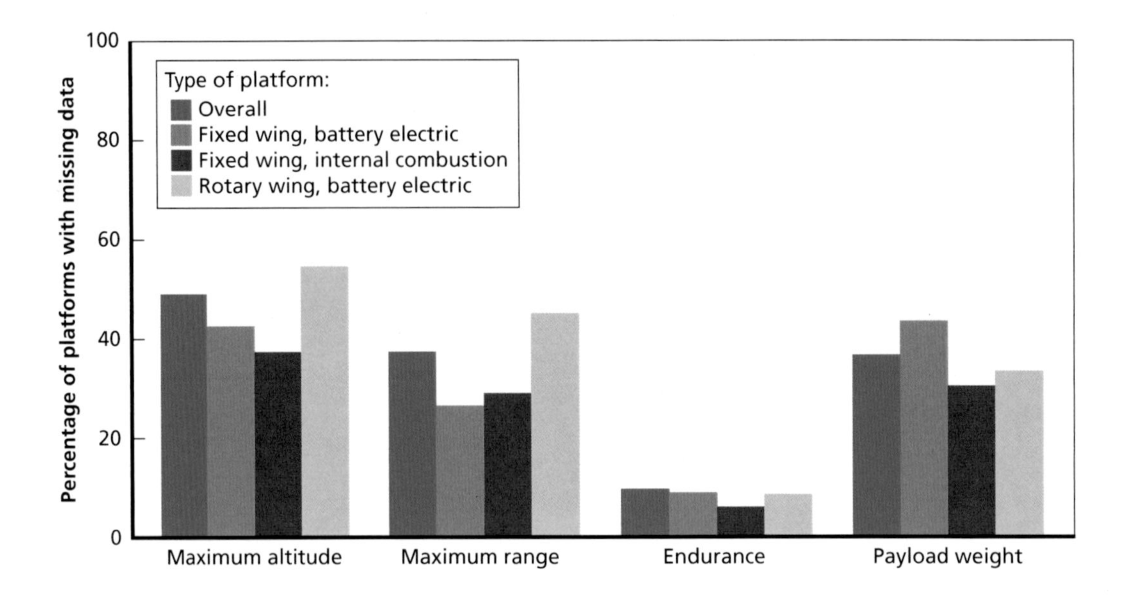

Figure 3.2
Missing Speed Metric Data in the Database

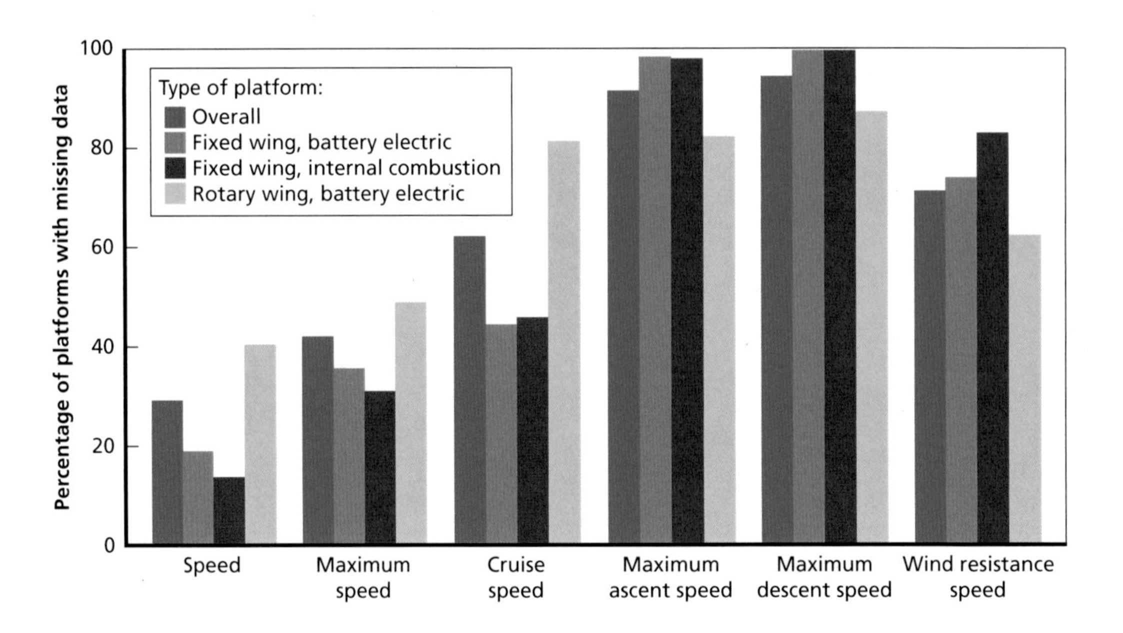

Figure 3.3
Types of Platforms in the Database

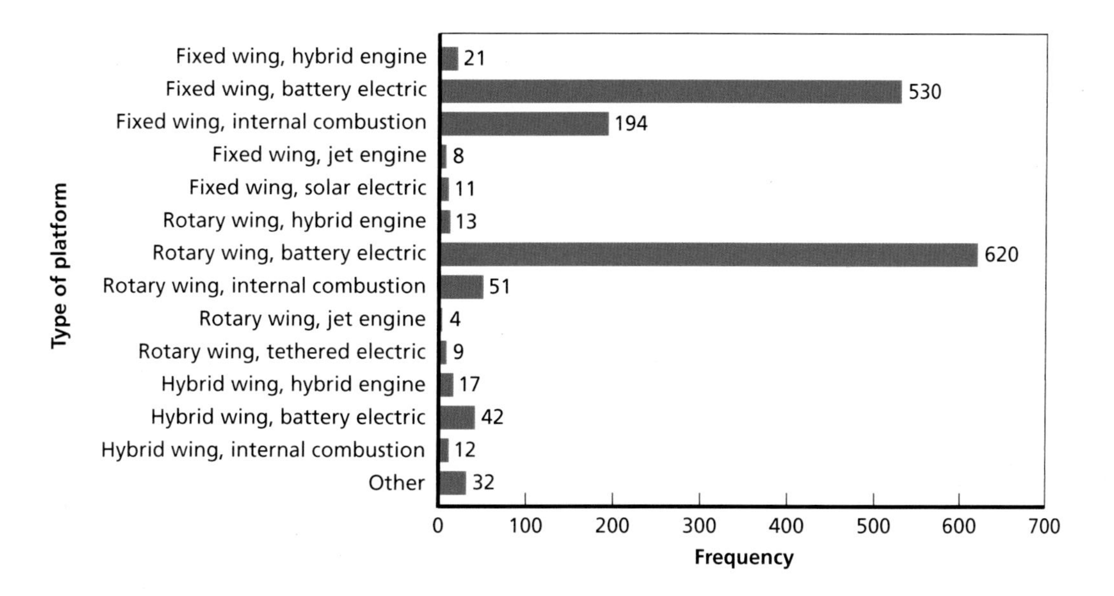

Figure 3.4
Maximum Range of Listed Models, by Platform Type

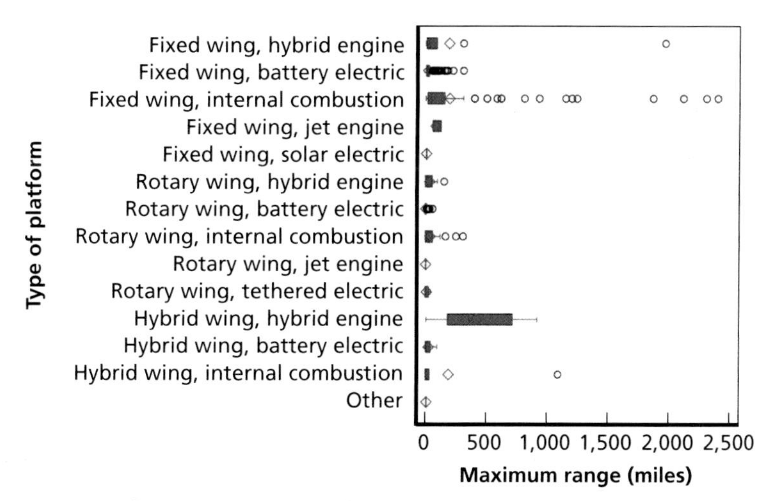

NOTE: The figure uses diamonds (mean value), blue bars (interquartile range [IQR]), a line bisecting the blue bar (median value), "whiskers" (indicating presence of data outside of IQR but within 1.5 × IQR), and circles (outliers, defined as more than 1.5 × IQR).

To better show the trade-offs in platform type, Figure 3.5 shows only those sUASs with range less than 400 miles.

We found substantial variation in the maximum range and endurance of the platforms in the database. On average, hybrid-wing/hybrid-engine models had the greatest maximum range (mean = 419 miles, median = 359 miles); however, several individual fixed-wing/hybrid and fixed-wing/internal combustion models had maximum ranges surpassing 1,000 miles. The majority of these models were intended for disaster response, environmental research or monitoring, imaging, surveillance, reconnaissance, observation, patrol and security, research, search and rescue, survey and mapping, and target acquisition. Figure 3.6 shows the endurance of the models by platform type.

To better show the trade-offs in platform type, Figure 3.7 shows only those sUASs with an endurance less than 24 hours.

Hybrid-wing/hybrid-engine platforms also had the greatest endurance on average (mean = 520 minutes, median = 390 minutes), followed by fixed-wing/hybrid engines (mean = 484 minutes, median = 168 minutes), fixed-wing/solar electric (mean = 420 minutes, median = 420 minutes), and fixed-wing/internal combustion engines (mean = 394 minutes, median = 180 minutes). Some individual platforms reported endurance values greater than 24 hours. These platforms were mostly fixed-wing/internal combustion engines; however, other types of platforms were also capable for

Figure 3.5
Maximum Range of Listed Models for Platforms with Maximum Ranges Under 400 Miles, by Platform Type

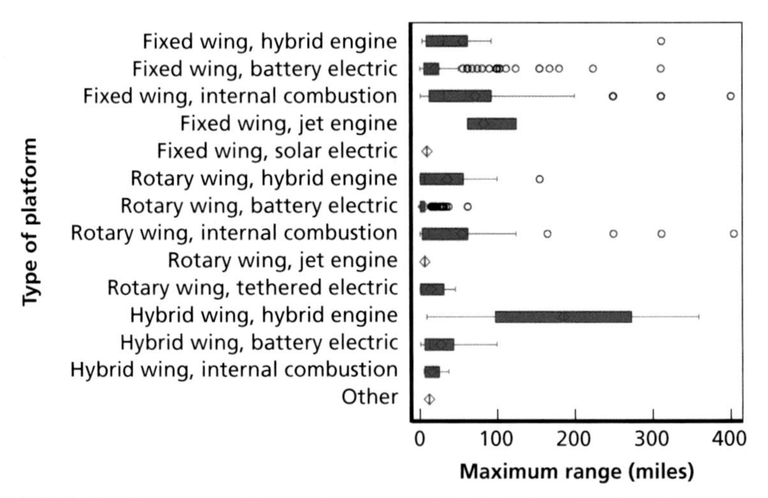

NOTE: The figure uses diamonds (mean value), blue bars (IQR), a line bisecting the blue bar (median value), "whiskers" (indicating presence of data outside of IQR but within 1.5 × IQR), and circles (outliers, defined as more than 1.5 × IQR).

Figure 3.6
Endurance of Listed Models, by Platform Type

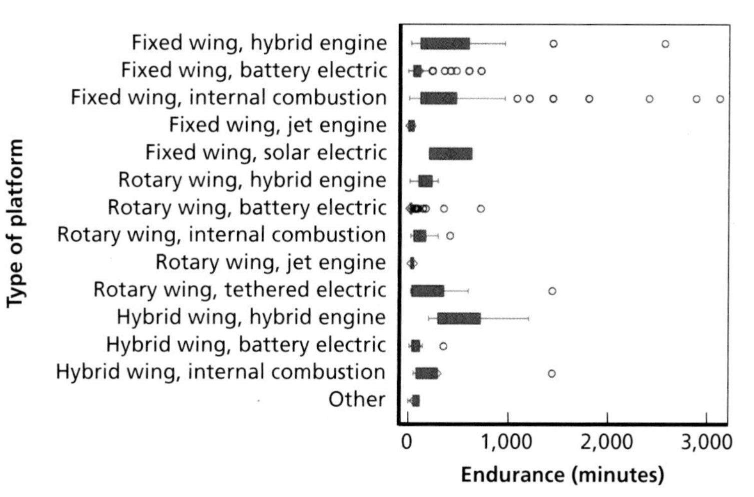

NOTE: The figure uses diamonds (mean value), blue bars (IQR), a line bisecting the blue bar (median value), "whiskers" (indicating presence of data outside of IQR but within 1.5 × IQR), and circles (outliers, defined as more than 1.5 × IQR).

Figure 3.7
Endurance of Listed Models for Platforms with Endurance Under 24 Hours, by Platform Type

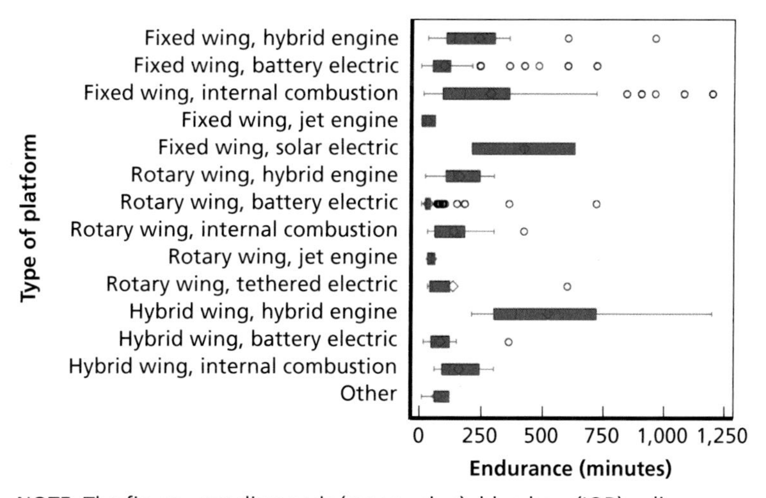

NOTE: The figure uses diamonds (mean value), blue bars (IQR), a line bisecting the blue bar (median value), "whiskers" (indicating presence of data outside of IQR but within 1.5 × IQR), and circles (outliers, defined as more than 1.5 × IQR).

flying for longer than 24 hours. Many of the platforms with endurance values of 24 or more hours were the same models with ranges of greater than 1,000 miles.

Payload Weight

We limited the sample of platforms to those with a MGTOW of less than 65 lbs.[1] Therefore, payload weight was also restricted (there were a few records in the database that had payload weights greater than the reported MGTOW). We modified the records of those that had documentation with the correct information and removed seven records from the database that we were unable to resolve.

Rotary-wing/jet engine models had the highest average payload weight; however, there were only two entries for this model type with non-missing payload weight fields. Rotary-wing/internal combustion engines had the next highest average payload weight (mean = 13.6 lbs, median = 11 lbs), followed by hybrid wing/internal combustion (mean = 12.3 lbs, median = 12 lbs), rotary-wing/hybrid electric (mean = 11.5 lbs, median = 15 lbs), and fixed-wing/internal combustion (mean = 10.7 lbs, median = 11 lbs), as detailed in Figure 3.8.

Figure 3.8
Payload Weight of Listed Models, by Platform Type

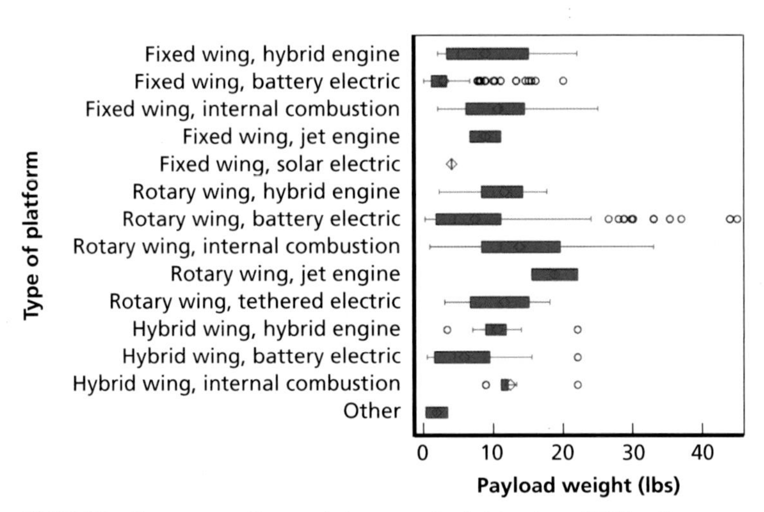

NOTE: The figure uses diamonds (mean value), blue bars (IQR), a line bisecting the blue bar (median value), "whiskers" (indicating presence of data outside of IQR but within 1.5 × IQR), and circles (outliers, defined as more than 1.5 × IQR).

[1] Some source data items were calculating MGTOW differently, and so the threshold for inclusion was set to 65 lbs rather than the typical 55-lb limit for sUASs.

While platforms with internal combustion or hybrid engines had higher average payload weights than battery-electric models, there were several individual battery-electric models with payload weights greater than 20 lbs. The models with greater payloads also had greater maximum range, maximum altitude, speed, and endurance. Payload weight was positively associated with maximum altitude, speed, and endurance after controlling for wing and propulsion type. There was no significant relationship between payload weight and maximum range after controlling for wing and propulsion types.

Maximum Speed

Unsurprisingly, we found that platforms with fixed-wing jet engine propulsion platforms have much higher speeds than other types of platforms, as shown in Figure 3.9. The next fastest sUASs, on average, were those platforms powered by internal combustion engines and hybrid engines, as shown in Figure 3.10.

However, there were several outliers: fixed-wing/battery electric platforms that reported speeds greater than the average for internal combustion platforms and one with a speed greater than some of the jet engine platforms.[2]

Figure 3.9
Maximum Speed of Listed Models, by Platform Type

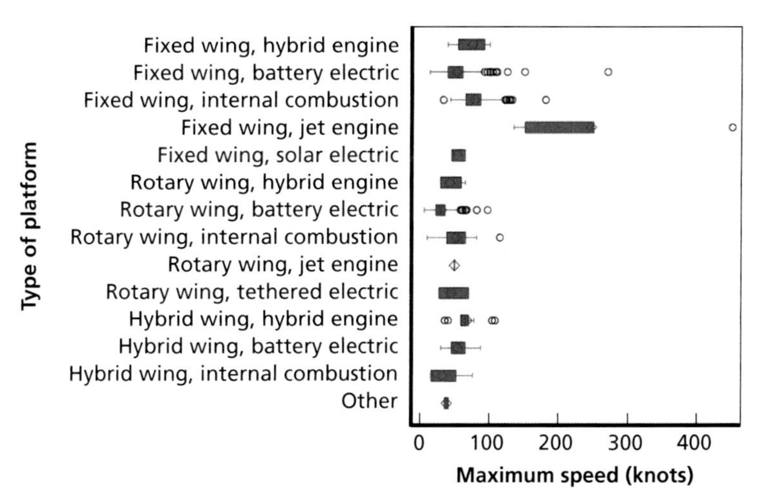

NOTE: The figure uses diamonds (mean value), blue bars (IQR), a line bisecting the blue bar (median value), "whiskers" (indicating presence of data outside of IQR but within 1.5 × IQR), and circles (outliers, defined as more than 1.5 × IQR).

[2] For example, the HAES 400 V1.6, an "autonomous aerial target used to provide a threat-representative target drone to support the Ground-to-Air Weapon System evaluation and test programs" (Esc Aerospace, 2012).

Figure 3.10
Maximum Speed of Listed Models, by Engine Type

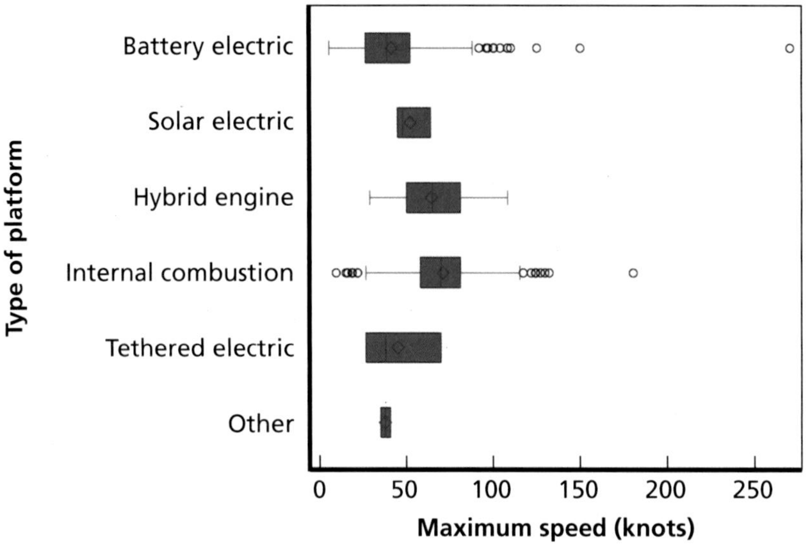

NOTE: The figure uses diamonds (mean value), blue bars (IQR), a line bisecting the blue bar (median value), "whiskers" (indicating presence of data outside of IQR but within 1.5 × IQR), and circles (outliers, defined as more than 1.5 × IQR).

Variation in Performance Metrics by Stated Intent

The data set included a field that described the "stated intent" of the models across 28 categories. We had data on stated intent for nearly every system (1,733 systems) in the subset, with at least one tag for each system. Figure 3.11 shows some of the more prominent classifications. Figure 3.12 shows the frequencies of some of the less common classifications, such as mining, pipeline services, and firefighting.

Initially, we hypothesized that prototypes would have better performance than other models in the database and that they might give us insight into the types of capabilities that are in development. However, using a linear regression model controlling for platform type, we found that the performance of prototypes according to each of the five performance metrics was not statistically different from that of other platforms. Ultimately, we dropped these models from our analysis, given that these prototypes were less likely to be commercially available.

Performance of Top Sellers Relative to Other Platforms

We used a similar approach to compare the performance of top-selling platforms to that of the rest of the field. Top-selling models are an important baseline of compari-

Figure 3.11
Commonly Cited Stated Intent Categories

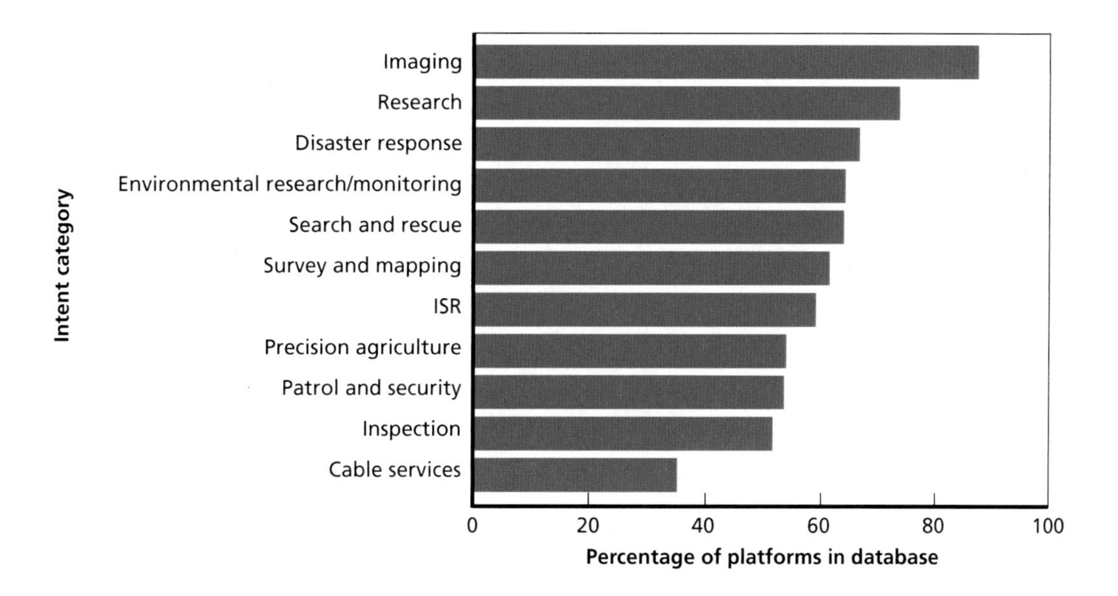

Figure 3.12
Less Commonly Cited Stated Intent Categories

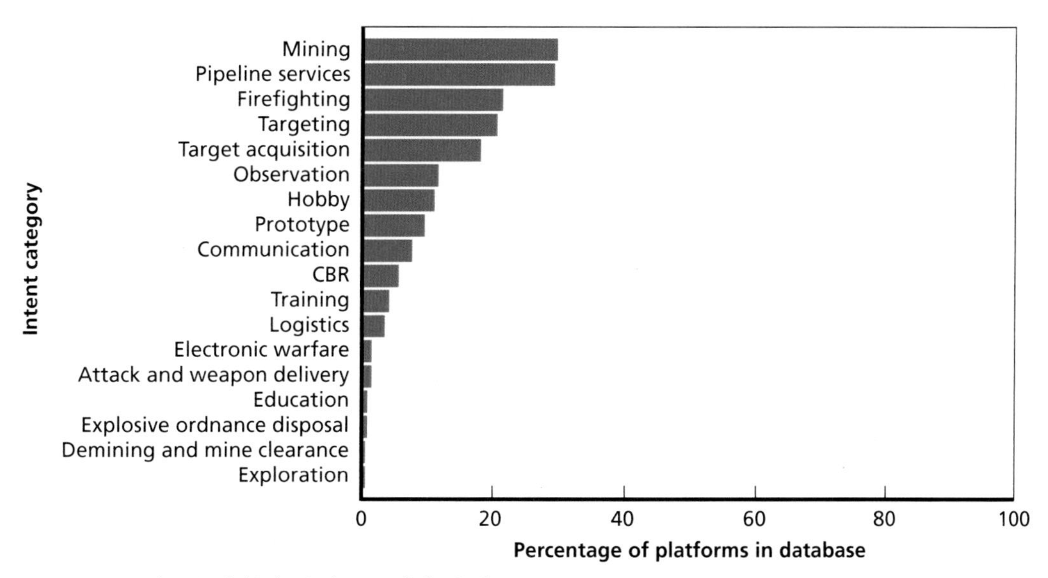

NOTE: CBR = chemical, biological, or radiological.

son because they are likely to be more widely available, have a larger user community, and would therefore be easier to tailor to a user's goals. Because most of top-selling platforms were in the rotary-wing/battery electric propulsion category, we limited this analysis to these types of platforms.

We found that top-selling platforms had lower-than-average endurance (mean = 21 minutes for top sellers, compared to a mean of 33 minutes for non–top sellers), lower-than-average maximum range (mean = 1.8 miles versus 4.7 miles), and lower-than-average payload weights (4.6 lbs versus 7.4 lbs). We found no differences in average speed or average maximum altitude.

Summary of Findings from the Performance Analysis

The performance characteristics have trade-offs. Increasing one capability comes at a cost to another: Increasing payload reduces endurance and reducing size limits sensor selection, all other things being constant. In examining these trade-offs, we found that fixed-wing sUASs—especially those with internal combustion engines—were the leaders in terms of range, endurance, and speed. However, rotorcraft can carry similar payloads, and their reduced detection signatures and ability to hover and perform agile maneuvers make them an attractive option for nefarious users.

In analyzing the trade-offs among performance characteristics, fixed-wing sUAS have range, speed, and endurance advantages over rotary-wing sUASs. However, hybrid designs that combine both methods of lift generation may offer a combination of improved range and endurance with improved maneuverability and station-keeping. Internal combustion engines offer improved power, leading to superior performance (e.g., speed, endurance, payload) but also increased complexity. In particular, there is increased mechanical complexity, a larger acoustic signature, and more complex interfaces with onboard electronics. Furthermore, internal combustion engines at very small sizes have diminished returns. These reasons explain why many manufacturers use battery electric power instead of internal combustion engines. This also implies that electric powered systems will very likely fall short of combustion engine systems in terms of range, endurance, and payload.

The analysis in this chapter points to several observations about sUAS trends and capabilities:

- There are many sUAS platforms on the market with unreported performance metrics, with maximum altitude and maximum ascent and descent speeds being some of the least well reported and potentially least understood.
- Rotary-wing battery-electric platforms represent 41 percent of sUASs, fixed-wing battery-electric platforms account for 35 percent, and fixed-wing internal combustion platforms make up 12 percent.

- On average, hybrid-wing hybrid-propulsion platforms have the longest range (419 miles) and longest endurance (520 minutes).
- Rotary-wing internal combustion platforms have the highest mean payload weight, at 13.6 lbs.
- The most commonly stated intents for sUASs are imaging, research, and disaster response.
- sUASs with a stated intent of prototype were not statistically better or worse than those with other stated intents in any of the five performance categories studied.
- Top-selling sUASs have lower-than-average endurance (mean of 21 minutes versus 33 minutes for non–top sellers), lower-than-average maximum ranges (mean of 1.8 miles versus 4.7 miles), and lower-than-average payload weights (4.6 lbs versus 7.4 lbs). We found no differences in average speed or average maximum altitude.
- Considering the increasing payload on performance, we found that fixed-wing sUASs—especially those with internal combustion engines—were the leaders in terms of range, endurance, and speed.

CHAPTER FOUR

Novel and Likely Nefarious Tactical Use Cases

To assess nefarious uses of sUASs, we first derived possible use cases from multiple sources. Primarily, we referenced use cases submitted for review by DHS components to the C-UAS Capabilities Analysis Working Group.[1] We sought out additional sources to augment these use cases. We conducted our own open-source research, informed by current events, to add possible use cases. These incidents cover a number of attack vectors and methodologies, including the non-autonomous kamikaze explosive swarming attacks on Russian bases in Syria in January 2018 (Binnie, 2018),[2] the reported harassment attack on Federal Bureau of Investigation agents in May 2018,[3] the July 2018 kamikaze explosive attack on a Mexican government official's home (Sullivan, Bunker, and Kuhn, 2018), and the August 2018 kamikaze explosive drone assassination attempt against Venezuelan President Nicolas Maduro (Binnie, 2018). We also identified examples of ranged explosive attacks enabled by newly released technology, such as a grenade launcher–equipped sUAS from the Ukrainian company Matrix UAS (UASweekly, 2018).

Because these uses do not span all target types and component missions, we sought to identify the threat usage categories that capture relevant uses of these technologies and identify a representative subset for case studies of high-impact scenarios. Our compilation of sources provides a baseline set of nefarious sUAS uses, which we extrapolated into more-generic categories of threats. To do so, we developed an sUAS adversary framework that captured the uses to the greatest extent possible and can be used to assess the likelihood and severity of types of adversary uses. This framework also structures these categories, or "threat vectors," hierarchically, as shown in Figure 4.1.

---

[1] The C-UAS Capabilities Analysis Working Group was established in 2018 by the DHS Joint Requirements Council and is composed of members of relevant DHS components.

[2] Multiple noncoordinated, automated sUASs concurrently converged on two Russian bases in Syria. Russian defenses managed to destroy or ground all of the UASs.

[3] Multiple human-controlled sUASs physically harassed agents during a hostage situation and streamed video of their activities online.

Figure 4.1
Adversary sUAS Framework

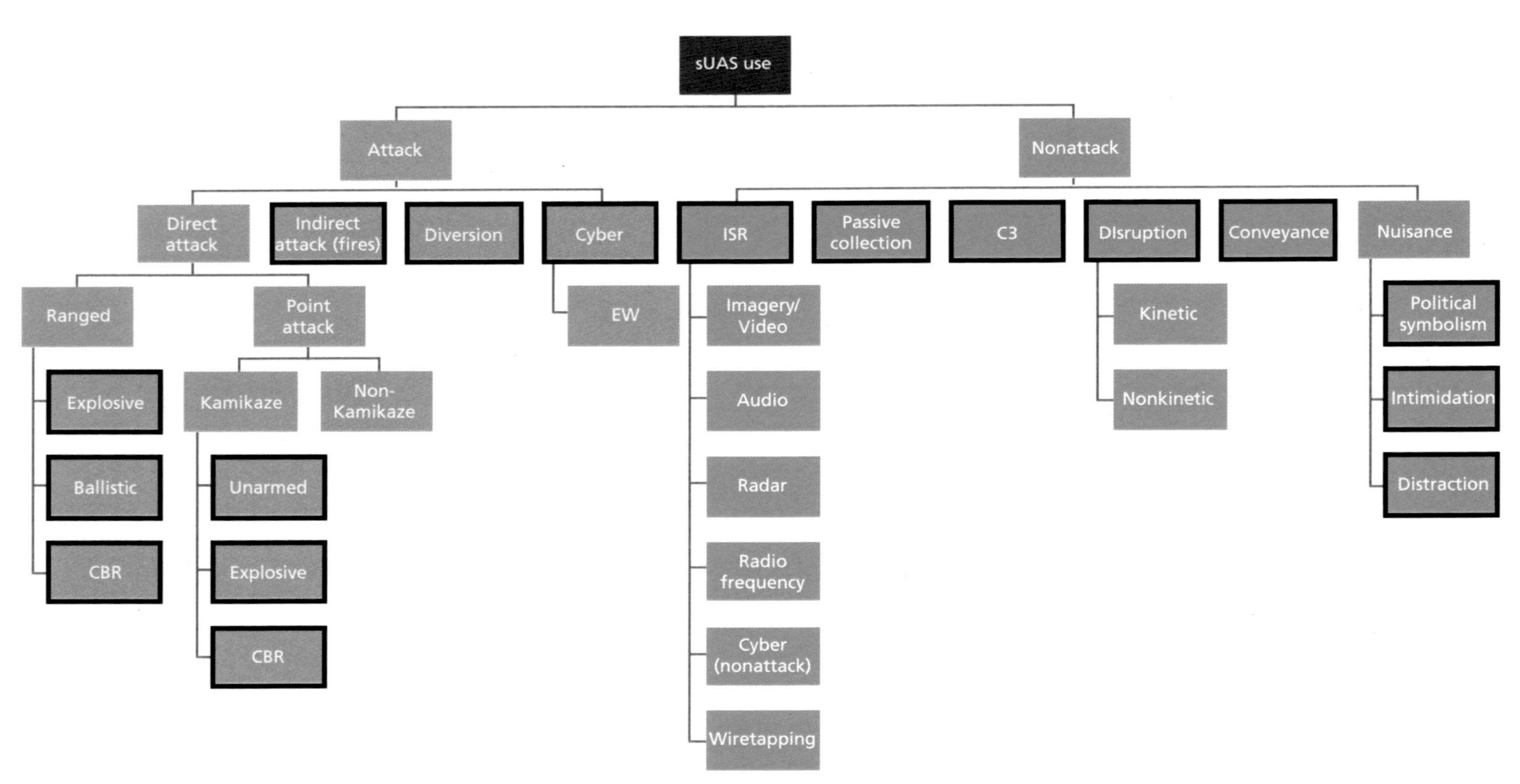

NOTE: Threat vectors highlighted in black are those that were used in the group risk analysis.

We broadly categorized adversary sUAS uses into *attack* and *nonattack*. In the attack category, we further decomposed the uses into *kinetic attack* and *nonkinetic attack*. In decomposing a kinetic direct attack, we focused on the type of weapon used (CBR, ballistic, explosive, unarmed) to allow a meaningful risk assessment of each category, in terms of both likelihood and severity.[4] The nonattack categories were a significantly more diverse subset, ranging from several types of ISR missions to conveyance (when an adversary uses a UAS to convey illicit items across or into restricted areas) and various types of uses categorized as nuisance (e.g., political symbolism, distraction, intimidation).

Using this framework, we sought to identify high-risk threat vectors for more detailed case study analysis. We outlined 16 threat vectors that covered the above framework[5] and used a multiround group elicitation exercise to assess risk as a study team. Table 4.1 describes the 16 threat vectors chosen for the assessment and provides an example for each. Each study team member and reviewer (11 HSOAC participants in total) rated the relative likelihood and consequence of each threat vector using 100 total "points" for each assessment.[6] After the initial rating, the participants were allowed to see results across all raters, discuss the results, and reassess.

Table 4.1
Threat Vectors

Threat Vector	Description	Example
Kamikaze point attack	Adversary directs UAS chassis into a target, with only the UAS itself used as a weapon	UAS intentionally flies into aircraft engine
Kamikaze point attack (with explosives)	Adversary directs UAS chassis into a target, with the UAS containing explosives for greater damage	UAS with plastic explosive lands on facility and detonates
CBR attack	Adversary uses UAS to launch a CBR attack, which could be kamikaze, spraying a substance, or firing a projectile from stand-off	UAS deposits dirty bomb to facility roof at night, and bomb activates the next day; UAS sprays aerosolized anthrax above a facility
Stand-off attack (with firearm)	Adversary uses UAS equipped with a firearm to engage a target	Uzi-mounted UAS shooting into crowd
Stand-off attack (with explosives)	Adversary uses UAS for delivery of explosives at range	UAS drops grenades into a crowd

[4] Note that an "unarmed kinetic attack" can also involve a UAS attempting to collide with another in an attempt to destroy it (or at least force it to abandon its mission).

[5] For simplicity, we did not always drill down to the lowest level; for example, we simply used *ISR*, rather than identifying each type of ISR mission in the framework. Additionally, we combined chemical, biological, and radiological into *CBR*.

[6] *Consequence* was undefined for the rater, but interpretations were discussed after the first-round scoring.

Table 4.1—Continued

Threat Vector	Description	Example
Indirect ranged attack	Adversary uses UAS for target acquisition and range-finding for a human-operated long-range weapon	UAS surveils a facility for soft targets to cue mortars located hundreds of meters away
Diversion in support of attack	Adversary uses UAS to distract or divert friendly forces in support of a larger, manned attack	UAS swarm draws security forces to far side of facility while manned attack hits main entrance
Active cyberattack/ disruption	Adversary uses UAS as a platform for other devices to launch malicious cyberattack	UAS uses location to gain local network access and installs malware that provides remote users access/ privileges
Passive electronic collection	Adversary uses UAS as a platform for other devices to collect electronic information from target	UAS with Wi-Fi sniffer lands on facility roof and monitors traffic; UAS captures two-way radio transmissions from law enforcement
C3 attack	Adversary uses UAS to support command of and communication between physically distant adversary actors	UAS serves a mobile relay node for line-of-sight–limited communications (including for other UAS); UASs track multiple smuggler operations for cartel boss
ISR	Adversary uses UAS to detect, identify, and monitor friendly forces	UAS finds and follows U.S. Border Patrol (USBP) agents to allow smugglers to evade them; UAS tracks security shift changes at facility
Disruption/ harassment	Adversary uses UAS flying in close proximity to friendly forces to hinder friendly operations	UAS swarm flies into law enforcement officers at outset of raid to buy adversary time; UAS buzzes aircraft during training exercise
Conveyance	Adversary uses UAS to convey illicit items across or into restricted areas	UAS transports drugs over border and into U.S. urban area; UAS transports barred weapons into prison
Political symbolism	Adversary uses UAS for act of political demonstration	UAS defaces hard-to-reach symbol at government facility; UAS spray-paints slogan on government vehicles
Intimidation	Adversary uses UAS in publicized demonstration of capability to elicit concessions from friendly actors	Massive UAS swarm used in coordinated show of force; UAS delivers supposed explosive payload and operators demand ransom
Distraction	Adversary uses UAS to distract friendly actors not in support of an attack	UAS swarm pesters lone USBP agent on foot; UAS loiters above facility courtyard

We intentionally did not describe a very specific use case for each vector because our objective was to cover the adversary problem space as comprehensively as possible. As a result, some of the vectors are open to the assessor's interpretation for severity (and likelihood). For example, trace amounts of a chemical agent delivered to a water source may not be nearly as severe as a large amount of chemical substances sprayed in a mass gathering. So, they may be assessed differently by different individuals.

On average, we rated the ISR and conveyance threat vectors as having the highest likelihood, which aligns with anecdotal evidence from the field. Many of the kinetic attack vectors were rated as low-likelihood, along with the "nuisance" threat vectors. Conversely, we assessed that the threat vectors with the most severe consequences were the kinetic attacks, particularly the CBR attack and the kamikaze explosive attack.

As stated earlier, the overall objective of this approach was not to definitively rank the threat vectors, but to identify a representative subset of threat vectors for case studies of high impact scenarios. Figure 4.2 plots the averages of the likelihood and consequence ratings in a scatter plot.[7]

Figure 4.2
Scatter Plot of Results

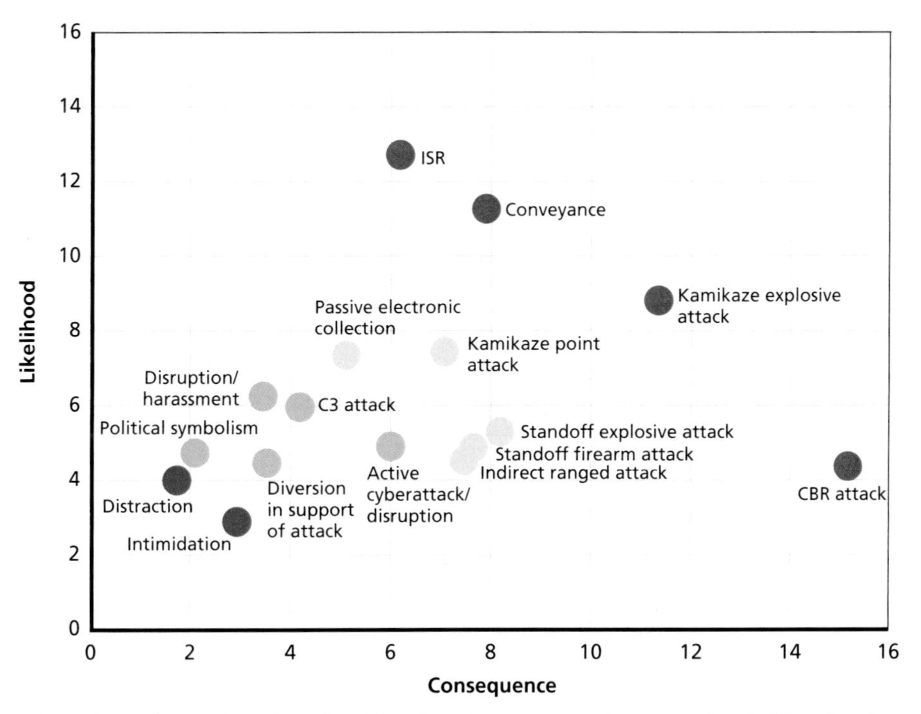

NOTE: The scales are based on the 100 points that raters used to score the likelihood and consequence.

[7] We used average risk scores (as opposed to median scores) despite the presence of outlier results because we believed that the outliers were still useful information here, or at least were no more likely to be wrong than other responses. Note that using the median scores still highlights these same four threat vectors as the riskiest.

There is a clear frontier of high-risk threat vectors ranging from high probability to high consequence:

- ISR
- conveyance
- kamikaze explosive attack
- CBR attack.

These four threat vectors will form the foundation of the high-risk use cases that we will examine in more detail in our case studies. We again note that the assessment described here was conducted with HSOAC team members and reviewers. Our objective is to have external subject-matter experts go through this exercise as well and to incorporate those results into our analysis.

Case Studies of Likely and High-Impact Scenarios

This analysis considered factors for each of the four red scenarios in Figure 4.2. Table 4.2 shows buckets for sUAS performance capabilities geared to meet the performance needs in Table 4.3. It is important to note that these values are exemplary. Table 4.3 highlights rough-order-of-magnitude performance characteristics needed for an sUAS to engage in a scenario.

There may be numerous vignettes within each threat vector that use values outside these relative ranges. Tables 4.2 and 4.3 highlight general parameters for an sUAS to perform a scenario within each threat vector. The intent is to demonstrate that there are sUASs on the market that are capable of carrying out these threats. The threat vectors are broad enough that additional sUASs beyond those noted in Table 4.3 might also be able to carry out these types of missions.

The following narratives describe four plausible instances of adversaries using sUAS technologies described in Chapters Two and Three, using the four likely, high-impact threat vectors described in this chapter to characterize the ends, ways, and

Table 4.2
Categories of sUAS Performance Capability

Performance Characteristic	Low	Moderate	High
Range (miles)	< 5	≤ 5, < 20	≤ 20
Endurance (minutes)	< 30	≤ 30, < 60	≤ 60
Payload (lbs)	< 2	≤ 2.2, < 10	≤ 10
Speed (knots)	< 25	≤ 25, < 75	≤ 75

Table 4.3
sUAS Performance Needs to Effectively Engage in Mission

					sUASs Capable of Completing Mission	
Scenario	**Range**	**Endurance**	**Payload**	**Speed**	**%**	**Number**
ISR	Low	High	Low	Low	23	332
Conveyance	High	High	High	Low	4	53
Kamikaze explosive attack	Low	Low	Moderate	High	5	72
CBR	Moderate	Moderate	High	Low	6	84

means by which UAS uses could threaten DHS's ability to carry out its missions. Each of these narratives describes a scenario that takes place two to ten years in the future, with sUAS capabilities based on realistic extrapolations from current market trends. Each of these scenarios is aligned with one of the four threat vectors (ISR, conveyance, kinetic kamikaze attack, CBR attack) rated to be of highest risk in our internal evaluations. Each scenario also aligns the mission sets of four DHS components, respectively:

- U.S. Coast Guard (USCG)
- USBP in U.S. Customs and Border Protection (CBP)
- Federal Protective Service (FPS) in DHS's Cybersecurity and Infrastructure Security Agency
- U.S. Secret Service (USSS) and DHS's Office of Operations Coordination (OPS).

Surveillance and Reconnaissance

There are approximately 330 systems available today that can meet the demands of the surveillance and reconnaissance use case. Seventy percent are fixed-wing platforms and 16 percent are rotary-wing platforms, with the balance being fixed-wing and rotary hybrids. Roughly half of the systems are fixed-wing battery electric, and a further 26 percent are fixed-wing internal combustion.

Notional Vignette: Human Trafficking ISR in the Caribbean

In the year 2022, the Bahamas remain a key transitway for human trafficking into the United States, despite ongoing efforts by the Bahamian government to combat it (U.S. Department of State, 2018). Traffickers use the regular boat traffic between smaller Floridian ports and various islands within the Bahamas as cover for their smuggling trips. Smugglers attempt to pose as benign fishing or pleasure vessels to evade authorities in both the Bahamas and the United States. As such, USCG Station Miami coordinates with the Bahamian government, among others, to try to intercept human traffickers by stopping and boarding suspicious vessels in the Florida straits. In response, trafficking gangs attempt to surveil USCG operations in the straits to evade the authorities.

One such group maintains a discreet surveillance outpost on the sparsely populated northernmost edge of Andros Island, northwest of Nicholls Town. The group recently acquired a state-of-the-art tilt-rotor UAS to supplement its manned surveillance. The platform is made of sleek carbon-fiber composite materials and weighs just 20 lbs (without a payload), despite being nearly three feet long from tip to tail. Most of the platform's weight comes from the fuel (battery and gasoline) and the hybrid internal combustion/electric engine, which provides approximately nine hours of endurance time.

The smugglers attach both an electro-optical/infrared (EO/IR) camera and an electronic intelligence (ELINT) receiver to the UAS. The EO/IR sensor records video at 4K resolution and pushes simultaneous 2K video to an operator on Andros Island via satellite. The higher-resolution video allows the onboard AI to leverage "follow-me" routing to locate and follow vessels with specific characteristics. The IR capabilities supplement the platform's ability to discern vessels in the open ocean. The ELINT receiver also allows the platform to locate vessels and to differentiate USCG and Bahamian government vessels from other vessels using direction and distance triangulation of specific radio signals. The group uses human sources in the Bahamian government to pass along information on the frequencies used by law enforcement and the USCG. Fully laden with the two sensors, the platform weighs approximately 32 lbs, which reduces its endurance to roughly six hours.

At the outset of a surveillance mission, the platform takes off vertically until it approaches its cruising altitude, a process that takes less than five minutes. The engine noise is audible, but the takeoff location is sufficiently remote that it does not get reported. As it approaches cruising altitude, the platform transitions its tilt-wings for forward movement and moves into the Florida straits. The movement of the platform is largely preprogrammed: The smugglers have placed waypoints along several popular smuggling routes between the Bahamian islands and the Florida coast, to which the platform travels in sequence. The first waypoint is at the Bimini Islands, about 75 miles west-northwest of the surveillance post and the closest Bahamian landmass to Miami. The second waypoint is halfway between North Bimini Island and Miami Beach (an additional 30 miles), and the third is three-quarters of the way between West End (on Grand Bahama) and West Palm Beach (a further 66 miles). See Figure 4.3 for more detail on this route.

As the platform flies along the set path, it continuously scans for vessels using both the EO/IR and ELINT sensors. A positive match on the correct RF induces the AI controlling the platform to diverge in the direction of the radio signal, looking for the corresponding vessel. The platform uses the IR sensors to increase the contrast of the vessel against the sea. The EO sensors are used to take numerous pictures of the vessel and then algorithmically compare them to known USCG vessels in the Station

Figure 4.3
Preprogrammed Flight Path for an Adversary Surveillance Tilt-Wing UAS (red) and Return Options (gray) Relative to USCG Station Miami (blue)

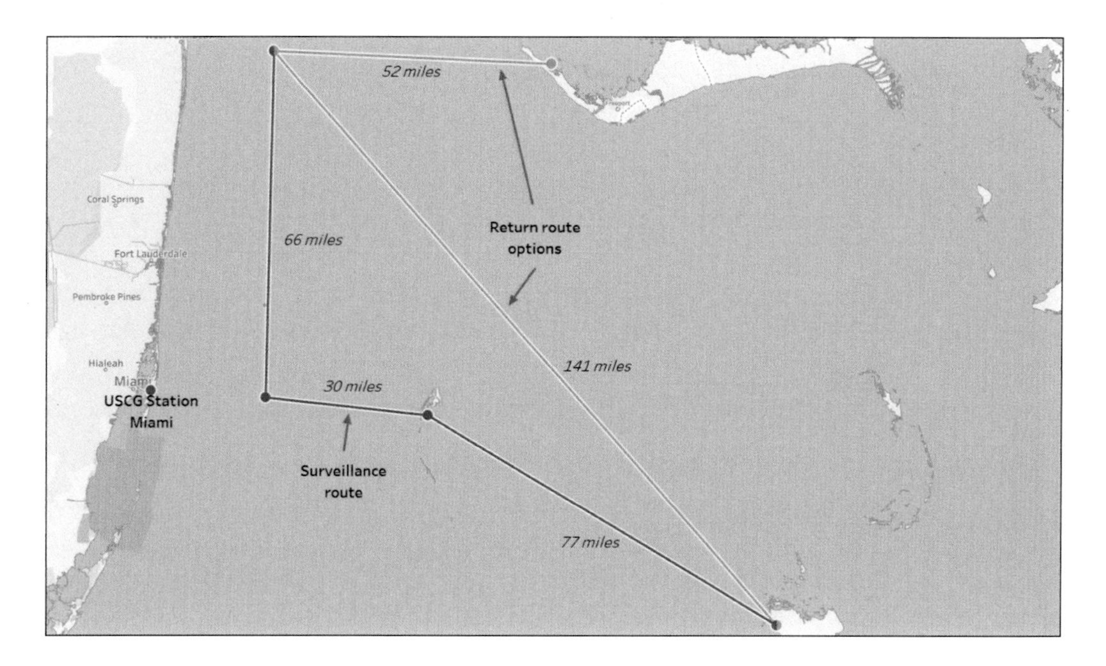

Miami fleet. With a positive match, the EO/IR sensors digitally zoom in on the vessel. The platform automatically shifts to a lower altitude and switches its wayfinding to "follow-me" to track the USCG vessel as it moves across the ocean. The platform also begins pushing a feed of the zoomed video to an operator at the Andros Island outpost. The operator can take copilot control over the platform to focus in on specific activities or to attempt to evade countersurveillance in the event that the platform is spotted. The information gathered will allow the smugglers to detect patterns in USCG patrol behavior and personnel activities. This will help the smugglers better evade detection when transiting from the Bahamas to points in Florida. Improvements in tactics based on UAS surveillance will help the cartels continue to illicitly move trafficked persons into the United States. The unimpeded pipeline will mean more demand to smuggle others into the United States, which has downstream effects on source countries in Central and South America.

After a successful surveillance operation, the platform can either continue on its waypoint-structured path or return to one of two bases of operation. The total distance of the surveillance route is 173 miles, which the platform can traverse in almost exactly two hours at its cruising speed of 75 knots. This gives the payload-laden platform sufficient endurance to perform between two and three 30-minute surveillance operations on a single launch. If the platform reaches the end of the surveillance route without spotting any target vessels, it will attempt to retrace its steps and fly back to the second waypoint parallel to the Florida coast. Once the platform has approximately two hours (or 173 miles) of endurance remaining, it will automatically attempt to return directly to one of the two landing sites operated by the smugglers: the surveillance outpost on Andros Island or a remote cove in the west of Grand Bahama, which serves as one of several staging points for smuggling vessels.

Conveyance

There are approximately 53 systems available today that can meet the demands of the conveyance scenario. This is surprising, given that rotary-wing systems are generally more amenable to higher payloads. However, the demand for a long range makes fixed-wing internal combustion and hybrid combustion the most common platform propulsion combination, accounting for 74 percent of sUASs currently manufactured. However, using a fixed-wing aircraft would limit delivery methods because it cannot hover or land and then takeoff as easily as rotary-wing platforms.

Notional Vignette: Drug Smuggling Conveyance at the U.S.-Mexico border

Drugs continue to flow across the U.S.-Mexico border in 2020 in the densely populated Tijuana River valley. Smugglers use a variety of means to move drugs over the border, including tunnels; vehicles, pedestrians, or off-road vehicles passing through ports of entry; and slow, low-flying aircraft. In the latter category are UAS platforms, which have only increased in use in 2020. Smugglers send UASs with significant drug payloads across the border on regular, preprogrammed routes, with minimal losses at this stage.

As an example of one of these routes, the Jalisco New Generation Cartel (which has successfully forced out the Tijuana and Sinaloa Cartels to control the lucrative Tijuana–San Diego drug route) operates a twice daily trafficking run between eastern Tijuana and San Diego. It does so using one of the latest DJI enterprise platforms (DJI Matrice 800 Pro). The Matrice 800 Pro, released in spring 2020, is a minor evolutionary upgrade of 2019's Matrice 800 (itself an upgrade of the Matrice 600 and 600 Pro). Like its predecessors, it is a rotary-wing platform with six rotors and a large base capable of carrying a sizable payload. The UAS has an entirely carbon-fiber frame and weighs 22 lbs (including its six batteries). It can carry up to an additional 11 lbs of payload at its base. One key improvement of the 800 Pro over its predecessors is a slight increase in endurance due to more efficient electric motors connected to marginally more compact batteries. As such, the platform can travel for roughly 24 minutes while fully laden. (Without a payload, the endurance is closer to 45 minutes.)

Each day, at roughly 05:00, the UAS takes off from a courtyard in the Rio Vista neighborhood of Tijuana carrying 10 lbs of processed heroin (street value: $435,000). The UAS ascends to 200 feet above the ground (in approximately 11 seconds), flies to a waypoint 2.5 miles east-northeast, changes direction, and quickly crosses the border heading west-northwest. This is to both avoid the airspace of the Tijuana Airport along the border and to use an unusual angle of attack to cross the border to confuse USBP and Mexican authorities about the actual origin point of the UAS. After traveling another two miles, the UAS hits another waypoint, shifts direction again (to the northeast), and travels another four miles to land at a predetermined location in the Otay County Open Space Preserve, where a cartel agent collects the platform and the payload and transports both to San Diego proper. In total, the UAS travels 8.5 miles in 17 minutes (an average speed of 30 mph).

The UAS is fully autonomous, and there is no camera on the platform (as a camera would add weight that could otherwise be used for drugs). The cartel changes the drop-off location and the waypoints regularly to keep the behavior of the UAS difficult to predict. The platform is not registered with the FAA or with local San Diego authorities.

The individual who collected the UAS removes the payload, refreshes the batteries, and launches the platform back to Tijuana on its second preset journey. It takes off from a junkyard near Chula Vista East in San Diego and heads nearly due south over the Pacific Gateway Park and across the border. Shortly after crossing the border, it cuts east for another two miles before returning to its original takeoff point. As before, this is both to avoid the Tijuana airport and to mask the UAS base of operations in Tijuana. Figure 4.4 provides more detail on both trips.

Figure 4.4
Preprogrammed Flight Path for an Adversary Drug-Conveying UAS (red) and Unladen Return Path (gray) Relative to CBP Otay Mesa Station (blue)

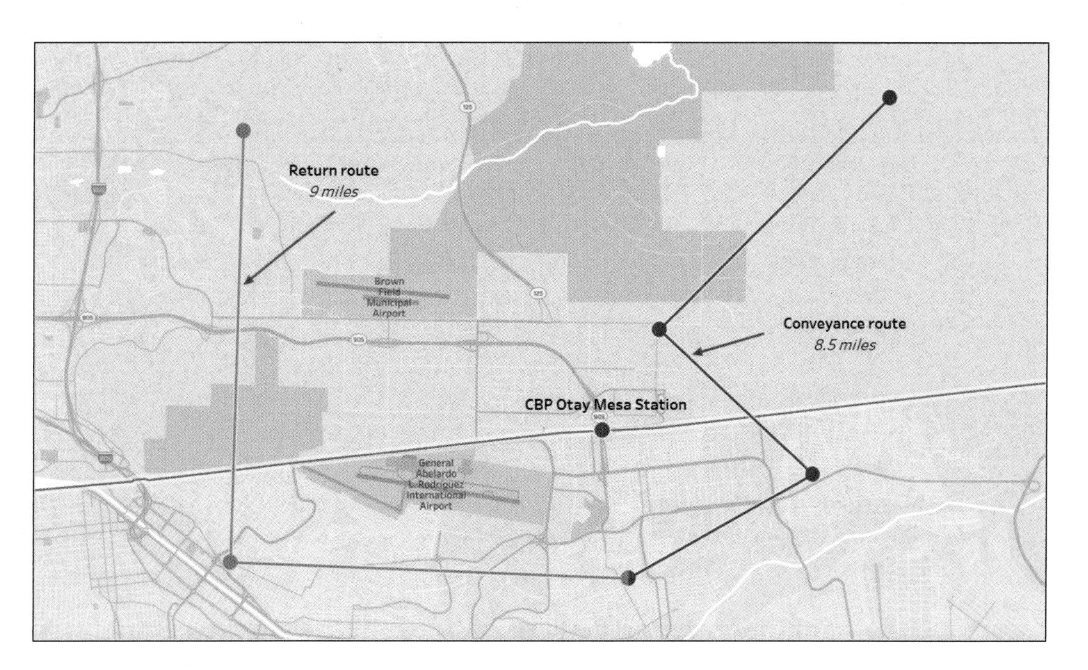

USBP agents at the Otay Mesa Station visually spot the UAS as it crosses the border from Mexico. Based on the pattern of the track it takes once spotted, they are able to identify the UAS as one operated by the cartel. However, USBP agents are unable to force the UAS to land or to intercept it at its destination due to the irregular movement.

Successfully returning to its home base in the Rio Visa neighborhood, the UAS batteries are recharged, and a new payload is packaged and attached. The platform performs a second, similar flight starting at roughly 20:30 that night. Completing both missions each day allows the cartel to move about $1 million in heroin across the border using a single UAS. However, the Jalisco New Generation Cartel operates multiple UAS drug conveyance units across the Tijuana River Valley, each with a similar concept of operations and payload. It is difficult for USBP agents and Mexican authorities to confront each of these threats.

Kamikaze Explosive Attack

There are few platforms (72 in production today) that can commit kamikaze attacks, as defined by the characteristics in Tables 4.2 and 4.3. However, the number would be substantially larger if speed is less of a concern. Available CUAS strategies and technologies for detecting, identifying, and disabling sUASs will dictate the role speed plays in this attack vector.

Notional Vignette: Kinetic Kamikaze Attack on Government Facility in Northern Virginia

FPS is responsible for responding to reports of unmanned aircraft approaching federal buildings and other facilities under its protection. According to its 2015 annual report,

> FPS views unmanned aircraft systems from a security perspective considering a full range of scenarios—from possible interference with law enforcement operations, to the potential for crashes and dangerous falling debris creating hazards for people and property, to the possible use of unmanned aerial vehicles for criminal payload deliveries or as a weapon of terror. Early on, FPS recognized the potential threat that could evolve from increased commercial and recreational use of unmanned aircraft and has actively contributed to the intelligence community's unmanned aerial vehicle threat working group and DHS and whole community preparedness working groups. (FPS, 2016, p. 36)

In fiscal year 2015, FPS responded to 19 individual incidents involving a UAS (FPS, 2016), and these incidents have only increased in frequency in the intervening ten years. By 2025, onsite FPS contract personnel have an established response procedure when a UAS is sighted near a federal facility. All sightings are routed to the local FPS Megacenter, which assesses UAS intent and provides additional guidance to the onsite personnel. FPS also regularly performs UAS risk assessments for all facilities under its protection.

In March 2025, Bob Hayward of Siler City, North Carolina, finds out that the Social Security Administration (SSA) has made its final decision not to recognize his back pain as a disability, and, thus, he is officially and irrevocably ineligible for disability assistance. Bob had been fighting with the SSA to recognize his back pain as a debilitating disability requiring disability benefits to supplement his part-time income.

Bob decides the SSA ought to feel some of the pain he has been through. Based on guides on the internet, he builds a crude nail bomb from fertilizer, household cleaners, and 3D-printed parts. He then pools his savings to buy the latest DJI midrange electric hexacopter on the internet. This platform weighs 10 lbs (with all batteries) and can fly for 15 minutes with a 5-lb payload (the approximate weight of Bob's bomb). Bob calls out sick from work and drives to Northern Virginia, bomb and UAS in tow.

The SSA Office of Disability Adjudication and Review is located in Skyline Tower in Falls Church, Virginia. It is a hard building to miss—perched on a hill and standing 24 stories tall, it dominates the surrounding skyline. Bob parks at Bancroft Park, a little under a mile away, and sets up his UAS, with the improvised explosive duct taped securely to the base of the platform beneath the camera. It is 14:00 on a cold, dreary April day and the park is deserted. Because Bancroft Park is within the 30-mile UAS-restricted zone around Reagan National Airport, Bob has physically disabled the GNSS on his device. He does not need the GNSS to navigate, as he plans to fly the device by camera alone, like he has practiced countless times at home. The device is registered with the FAA, but to Bob's neighbor Phyllis (who Bob hopes will take the blame for what is about to happen).

With Bob piloting the device remotely using the linked controller (which displays on its eight-inch monitor a 2K-resolution feed pushed from the UAS camera), the platform shakily ascends from Bancroft Park. It ascends to roughly 50 feet in ten seconds and moves in the direction of Skyline Tower. The device is high enough that the whir of the engines is difficult to hear from the ground but not so high as to be easily spotted from afar. The tower is 4,500 feet away from Bob's launch point, which means it reaches its destination in four minutes when traveling at its top speed of 15 mph (see Figure 4.5). With a maximum communications distance of 30,000 feet, Bob can move freely while controlling the device. He walks briskly back to his car. Even though Bancroft Park is still deserted, he does not want to be in the vicinity during the aftermath of the explosion. When the UAS is close enough to the building to nearly touch the tenth-floor window, Bob triggers the bomb from his cell phone and drives off.

An FPS security guard onsite notices the UAS as it crosses Leesburg Pike. Following procedure, she quickly reports the sighting to the FPS MegaCenter in Suitland, Maryland, and notes that the UAS seems to be making a beeline for the building. Unfortunately, 15 seconds after the UAS is spotted, it detonates. The force of the explosion shatters windows on floors 9, 10, and 11 of Skyline Tower. Several people are

Figure 4.5
Approximation of the Flight Path of Bob Hayward's UAS from Bancroft Park Takeoff to Social Security Administration Offices in Skyline Tower

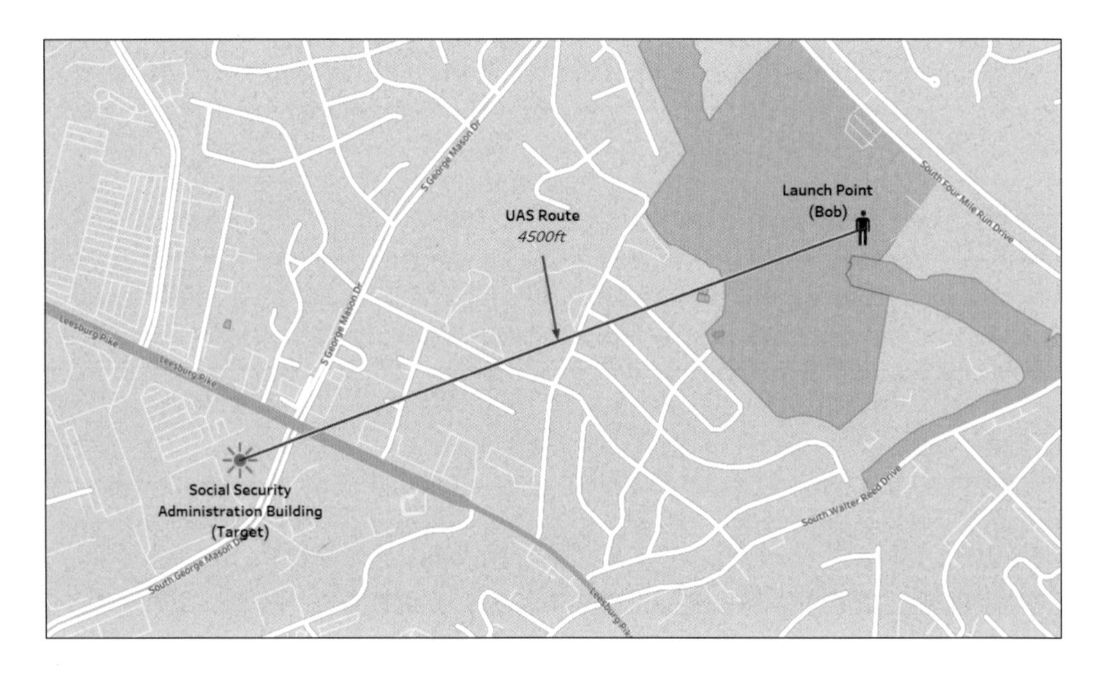

severely wounded from glass shards and projectiles embedded in the bomb. FPS personnel help secure the scene and triage the wounded until local police and emergency personnel arrive four minutes after the explosion. According to eyewitnesses who saw the UAS take off from Bancroft Park, Four Mile Drive and two consecutive exits onto Interstate 395 are immediately closed down. Unfortunately, by then, Bob has already made it to the interstate and is speeding south toward North Carolina. Only later would forensics identify the him as the culprit. Though no one died in the incident, some of the injured are permanently disabled. Thousands of federal workers call out sick or change jobs in the following weeks out of fear of copycat attacks.

Chemical, Biological, and Radiological

The field of systems capable of carrying out a notional aerosolized attack of a chemical, biological, or radiological agent, which requires a payload of greater than 10 lbs, the ability to fly for at least 30 minutes, and a range of at least five miles, is mixed. There are about 40 fixed-wing and 40 rotary-wing systems that exist today that can complete such a mission. If the retail commercial market is considered, then there are eight fixed-wing and five rotary-wing sUASs with this capability.

In this scenario, the assumption is that there is an attempt to disperse some sort of toxic agent over a wide area.[8] Therefore, a moderate amount of range and endurance is necessary to reach the target area and slowly loiter over it, as well as the capability to carry a high payload and maintain a low speed.

Notional Vignette: Chemical Attack on a Special Event in Southwest Florida

The Secretary of Homeland Security can designate any event as a national security special event (NSSE) at the request of the state's governor, provided the event is sufficiently high-profile to attract a potential national security threat. In making an NSSE designation, the Secretary considers many characteristics of the event, including

- anticipated attendance by U.S. officials and foreign dignitaries;
- size of the event; and
- significance of the event. (Reese, 2017, p. 2)

Similar to an NSSE are events assessed against the Special Event Assessment Rating (SEAR), which ranges from 1 (highest risk) to 5 (lowest risk). SEAR 1 and 2 events receive substantial federal support to provide capabilities and additional capacity that local and state law enforcement lack, potentially including C-UAS capabilities. DHS OPS assesses events submitted by state and local governments against the SEAR criteria to determine where events fit along the SEAR spectrum. For SEAR 1 and 2 events, the Secretary designates a federal coordinator for the event, who serves as the interface between the federal government and state or local governments in coordinating protection for the event (McLees, 2014). Multiple DHS components can play a role in the federal support provided to SEAR 1 and 2 events, including USSS, CBP's Air and Marine Operations, USCG, Transportation Security Administration, Federal Emergency Management Agency, and Immigration and Customs Enforcement's Homeland Security Investigations (Hosenball, 2015).

The yearly Super Bowl, as one of the largest sporting events in the United States, is considered a SEAR 1 event (McLees, 2014). In 2028, Super Bowl LXII is no different: OPS designated it a SEAR 1 event in November 2027 (ahead of the scheduled February 6, 2028, event). The Secretary of Homeland Security appointed a USSS special agent as the federal coordinator. USSS and other DHS components have surged personnel and equipment into and around Raymond James Stadium in Tampa Bay, Florida (the host of the event), in coordination with the Florida Department of Law Enforcement, Hillsborough County Sheriff's Department, and Tampa Police Department to ensure a common response if there is an attack.

Two years earlier, in retaliation for U.S. military support for the Algerian and Malian governments, a sophisticated cell of Al Qaeda in the Islamic Maghreb (AQIM) begins plotting against the event. The group acquires a small fleet of UASs: Three DJI

[8] This could be an aerosolized or sprayed agent dispersed by the UAS or by a device dropped by the UAS.

agricultural sprayer octocopters (successors systems to the DJI MG-1 and MG-1S) and 24 smaller quadcopter DJI Phantom 6 Pros (first released in 2025). Members filled each of the octocopters with half a gallon of strong neurotoxin pesticide phenylsilatrane (adding about 5 lbs to the weight of each UAS). The smaller quadcopters have only the small built-in camera as their payload.

The AQIM cell plans to spray the neurotoxin on the crowd from the octocopters while the quadcopters swarm around the octocopters, eight quadcopters per octocopter. The intent of the swarm is to add to the chaos and confusion during the attack while providing protection to the octocopter from C-UAS measures that AQIM anticipates DHS will have in place to secure the event. The swarm will be controlled entirely by the internal AI, with each quadcopter programmed to circle the central octocopter while coordinating on the fly with the other seven platforms in the swarm. Each device pushes a precise three-dimensional position relative to the octocopter, as well as velocity readings, 30 times per second over Bluetooth to the others in the swarm to ensure that the platforms do not crash into each other. Position relative to the octocopter is determined algorithmically based on the camera feed, and triangulation relies on a UHF signal transmitted from the octocopter. The octocopters fly using waypoints from the three separate launch points to converge on the stadium, supported by simultaneous localization and mapping (referencing a preprogrammed 3D map of Tampa) to avoid obstacles and optimize routing. The operators can manually control the devices using MOMU technologies. The octocopters can push 4K video feeds back to the operators and communicate over a spread spectrum. Finally, the octocopters have a kill switch to empty their tanks of pesticide immediately if they ever lose contact with the operators.

The operators also plan to hijack registered civilian UASs to add to the chaos around the stadium. A vehicle masquerading as a news van will push a signal that mimics that from legitimate air traffic management systems. It will override the routing of nearby UASs registered with the FAA and instead send hijacked platforms on a circuit circling the stadium 200 feet out at a height of 20 feet until they run out of fuel. The intent is to confuse and overwhelm any C-UAS defenses while attracting more of a crowd into the path of the attack.

The attack begins at the end of the fourth quarter as the largest group of fans exits the stadium, exuberant that the New York Giants have won their fifth Super Bowl ring. At this moment, the three separate octocopters lift off from their launch points from vans parked east of the Hillsborough River (the unofficial boundary of the USSS cordon protecting the stadium), followed shortly by each octocopter's accompanying quadcopter swarm, and converge on the stadium. Figure 4.6 shows more detail.

The nine-platform octocopter-quadcopter swarm nearly due east of the stadium lifts off from Duran Park, immediately crosses the Hillsborough River, and largely follows West Tampa Bay Boulevard directly to the stadium. Traveling at its top laden speed of 40 mph, the swarm reaches the stadium in just over two minutes. Simultaneous, hijacked UASs from near the stadium begin to converge and circle. The rapid

Figure 4.6
Preprogrammed Flight Paths of Three Separate UAS Swarms (red) Converging on Raymond James Stadium (blue) After the Super Bowl

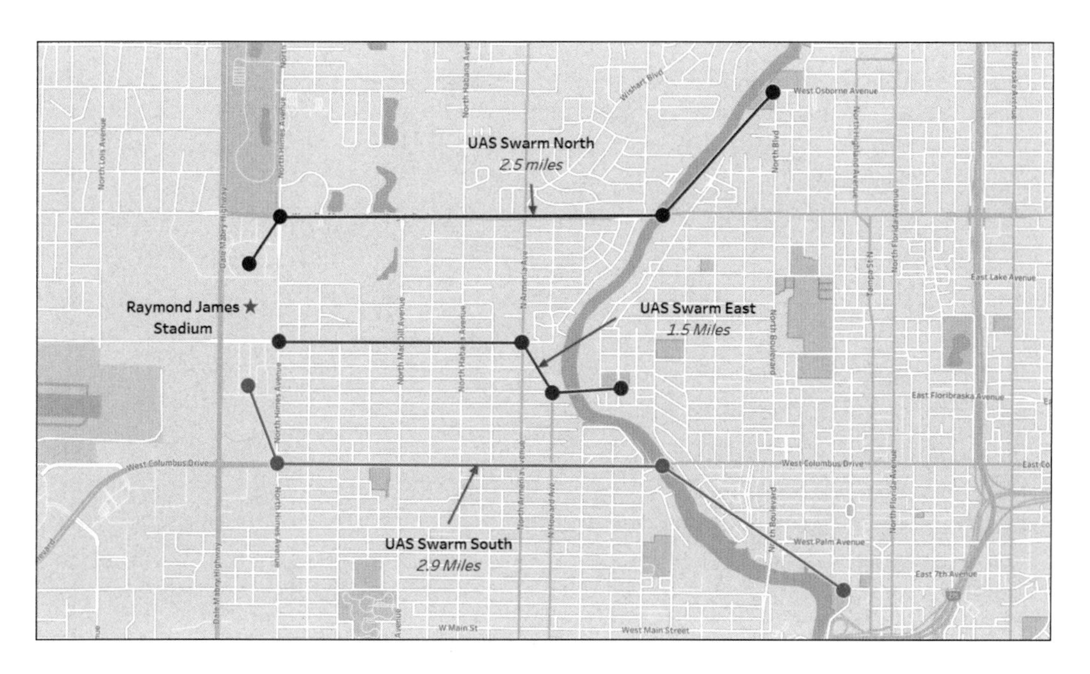

convergence of multiple UASs, including many of those that were cleared as non-threats to record the celebrations outside the stadium, confuses C-UAS personnel and delays the recognition of the octocopter-quadcopter swarm as the real threat. Once the swarm reaches its preprogrammed destination, the operators take over the octocopter and begin spraying the payload onto the crowd from a height of 25 feet, slowly moving the octocopter (and the rest of the swarm with it) in small circles to cover more area. Phenylsilatrane has a delayed effect in larger mammals, and the danger of the cloud of aerosolized liquid fogging down is not immediately clear to the crowd. However, the noise, gas, and sight of hijacked UASs clumsily crashing into obstacles on their crudely preprogrammed routes is enough to start a small panic in the crowd, shepherding people away from the stadium and toward the parking lots.

The octocopters stationed north and south of the stadium take off at the same time as the one to the east, but they have a greater distance to fly and thus take longer to reach the destination (see Figure 4.6). To the north, an octocopter takes off from Riverside Park (alongside its accompanying quadcopter swarm), travels southwest along Hillsborough River until reaching West Martin Luther King Boulevard, and follows this road west until reaching the park directly north of the stadium (a total distance of 2.5 miles). To the south, the octocopter-quadcopter swarm takes off from Water Works Park traveling northwest along the river until reaching West Columbus Drive, which it follows west to the parking lots directly south of the stadium (total distance traveled: 2.9 miles). The north UAS swarm reaches its destination about four minutes after takeoff (roughly two minutes later than the first swarm). The south swarm reaches its destination roughly 4.5 minutes after takeoff (2.5 minutes later than the first swarm). As the north and south swarms reach their destinations, the crowd has begun to disperse from near the stadium into the park (to the north) and parking lots (to the south), providing more targets for these two swarms. Using MOMU capabilities, the operators take control of all three octocopters and direct the north and south octocopters to begin spraying their targets.

The appearance of these two additional swarms, each spraying a noxious cloud like the first swarm, causes true panic to set in. Spectators and celebrants begin to stampede in every direction, trampling others underfoot. The thick cloud of pesticides in the air makes breathing difficult, further panicking the crowd. Within minutes, USSS personnel disable the octocopters, which triggers the kill switch to immediately pump out the remaining unsprayed pesticide. Local police attempt to enforce order amid the chaos and evacuate as many people as possible. Soon after, the pesticide (a convulsant) begins to take effect, including on first responders, further complicating the immediate response. In the end, dozens die and hundreds more report injuries. Hundreds of subsequent mass-gathering events for 2028 are cancelled across the country, costing billions in lost revenues. As a political statement, the attack does not succeed, as the U.S. policy toward Algeria and Mali does not change. However, as propaganda, it succeeds wildly, with videos from the event going viral on social media and thousands of recruits from across the world rallying to join AQIM's cause in northern Mali.

The AQIM terrorists attempt to flee the scene once their octocopters are disabled. The three vans quickly join the traffic on Interstate 275, two heading north and one heading south. The two traveling north are pulled over on Interstate 75 thanks to coordination between OPS, USSS, and Florida Highway Patrol. The van that traveled south is still unaccounted for.

Modeling sUAS Attacks

All these scenarios envision avenues in which adversaries could potentially undermine DHS missions using sUAS capabilities discussed in Chapters Two and Three and along the threat vectors discussed earlier in Chapter Four.

Additional modeling and simulation predicated on scenarios like these could help inform changes to DHS authorities, policies, and capabilities to enable the organization to better confront these threats. The remainder of this section provides examples.

Modeling and Simulation to Inform Time to Respond

One common element across each of the scenarios was the speed of current and near-future sUAS platforms (even when laden with fuel and a payload). Boosted by a nearly unrestricted set of launch points, the attacks provided DHS responders with very little time to work through the detect-classify-identify-respond kill chain. In these scenarios, UASs could appear and achieve their objective within minutes, even when launched miles from their destination. This observation echoes the conclusions of the USSS in response to an April 2014 incident in which reports of a small aircraft penetrating the Washington, D.C., airspace were still working their way through the chain of command by the time a gyrocopter reached the U.S. Capitol (Clancy, 2014).

Modeling different adversary concepts of operations (CONOPS), UAS platforms, and zones of operation can identify a distribution of realistic response windows, given specific mission sets and adversary objectives. Simulations of different DHS detection and response CONOPS for those mission sets and adversary objectives can identify the response CONOPS that DHS can effectively complete in time to combat adversary UAS usage. This analysis can also help determine the necessary detection ranges for different UAS platforms.

Modeling and Simulation for UAS Detection Capabilities

Another commonality across scenarios was that the adversaries tended to program or operate their sUASs to fly in a manner inconvenient for detection through traditional means. In most of the scenarios, the UAS had a small radar cross-section (RCS) (smaller UASs were sufficient to carry the needed payload) and flew low and sometimes fast through the built or natural environment. They also made quick, sharp changes in direction and altitude. Traditional radar systems, built to detect and track large, fast-moving, high-flying aircraft, are ill suited for tracking the platforms envisioned in these scenarios. Chapter Five addresses this issue in more detail and proposes radars with wider bandwidths, advanced digital signal processing, and frequency and waveform agility to better detect these types of UAS platforms. Additional modeling could explore the potential capabilities and limits of these technologies to track UASs, given the CONOPS articulated in these scenarios. Simulations could then compare the effectiveness of these modified radars against more novel detection means (e.g., acoustic sensors) and traditional EO/IR sensors (including human spotters).[9] We note that scenarios such as surveillance and reconnaissance and, possibly, CBR will typically have higher probabilities of detection because they spend more time within the defender's surveillance footprint.

Modeling and Simulation for Hostile sUAS Identification

One of the challenges raised by these scenarios, especially in those set further in the future, was the integration of innocuous sUAS use in the response environment. This trend will put a significant increased burden on the identification of hostile UASs and

[9] Other important factors that determine sUAS detectability include shape, material composition, and emissions (electronic, audio, IR), as well as the environmental conditions in which they are operating.

will increase the need for precise response tactics. Machine learning and other pattern-recognition tools could aid human observers in differentiating hostile from innocuous behavior. Additional modeling could also compare the precision of various response technologies (e.g., area of effects). Finally, simulations could test the effectiveness of these tools in reducing damage to innocuous UAS traffic while effectively defeating the adversary UAS.

Summary of Findings from Nefarious Use Case Analysis

Fixed-wing systems provide the necessary speed, range, and endurance but mostly lack rotary-wing systems' low speed, hover, and VTOL capabilities. Both types of platforms may be used in scenarios like the ones presented here, but multicopter rotary-wing platforms offer significantly lower training thresholds. Furthermore, internal combustion engines tend to be better suited to conveyance, kamikaze, and CBR attacks, but in choosing this option, attackers would give up stealth due to the noise of the engine compared with battery-electric propulsion.

We developed an sUAS adversary framework capturing 16 threat vectors, and then assessed the likelihood and severity of each vector relative to each other. In the process, we made the following observations:

- Systems exist today that can meet the requirements of scenarios that are either highly likely, such as ISR and conveyance, or highly consequential, such as kamikaze attacks or aerosolized CBR agent spreading.
- C-UAS strategies and technologies will have to account for both fixed- and rotary-wing aircraft powered by internal combustion and battery-electric propulsion across the range of DHS targets.
- The sUAS threat vectors of ISR, conveyance, kamikaze explosive, and chemical, biological, or radiological attack emerged as the most likely and consequential types of threats.

CHAPTER FIVE

sUAS Detectability

The ability to detect an sUAS depends on the sensors doing the detecting, and each of these sensors has trade-offs. sUASs present unique challenges to C-UAS systems due to their size, speed, and operational environment, and detecting, classifying, identifying, and tracking sUASs often requires numerous sensors working in concert.

Furthermore, nefarious sUAS operators will seek to reduce their devices' detectability along various detection axes. Even lawful operators may seek to reduce their signature to avoid becoming a nuisance. Lawful operators seek camouflage or reduced size to limit visual sensing or mask acoustic signature, which can provide advantages to nefarious operators as well. While reduced RCS and command-and-control signal obfuscation may not be explicitly sought by lawful operators (and therefore market forces), they may appear as byproducts of other developments (e.g., airframe material or aircraft design and miniaturization affecting RCS or cell signal control unintentionally obfuscating operators).

There is an array of sensor modalities for detecting, tracking, and identifying counter-UAS capabilities, including radar, EO/IR, ELINT, acoustics, and LIDAR. Each modality offers a range of capabilities with its own strengths and limitations, depending on the individual sensor. Combining multiple sensors/modalities into a system-of-systems may offer the most effective approach to countering UAS threats:

- Some sensors have wide fields of view and can search large volumes of space relatively quickly. These tend to be lower-resolution sensors that are ideal for initially detecting targets but not necessarily for tracking, characterizing, and identifying them. Examples include lower-frequency radars, ELINT, and acoustic sensors.[1]
- Others, like higher-bandwidth/frequency radars, have both the angular and range resolutions to track multiple targets within one to two degrees in bearing and tens of meters in range.

[1] It is possible for ELINT and acoustic sensors to identify targets if the target signatures can be accurately matched with those in a predefined database.

- Finally, high-resolution imaging sensors that have very narrow fields of regard can actually identify a target within a certain range.[2]

An effective systems-of-systems approach would be to search for and detect targets of interest using wide-area/lower-resolution sensors and then cue higher-resolution sensors to track and identify them. This requires a high degree of timely coordination among the various sensors.

It should be noted that all sensors and associated systems are subject to false alarms, both internally generated and external to the sensor, in the form of birds, wireless emitters, rain, or other sources of interference.[3] To reduce the number of false alarms, most sensors can raise their detection thresholds, thereby rejecting signatures below the threshold. However, this is problematic against small targets, as an increase in the detection threshold may lead to the rejection of nefarious sUASs as well. This can play havoc on the human operators monitoring the sensors and systems, desensitizing them to real threats and further limiting overall system effectiveness. In considering C-UAS systems and operations, one should keep in mind the false alarm rate of a given system, particularly from nuisance false alarms, and the impact on human performance in terms of monitoring and reacting to system outputs.

In this chapter, we discuss different modalities—radar, EO/IR, ELINT, acoustic, and LIDAR—and the key performance parameters that determine their capabilities against sUAS. In some cases, we describe and evaluate selected sensors in the C-UAS mission.

Radar

Radars emit energy in the form of electromagnetic radiation which propagates through space as electromagnetic (EM) waves. The performance of a radar system can be judged by the following: (1) the maximum range at which it can see a target of a specified size, (2) the accuracy with which it can determine the target's location in terms of range and angle, and (3) its ability to detect the desired target echo when masked by large clutter echoes, such as buildings or trees. These capabilities are either stated explicitly or contained within a set of standard radar performance parameters, as described next. When selecting a radar, one must consider the mission or set of missions to be performed, as performance requirements may vary across missions.

---

[2] There is a typical trade-off between sensor resolution and its field of regard or, more precisely, the size of the area it can effectively search within a given time.

[3] External false alarms are also referred to as nuisance *false alarms*.

- *Detection range* is the maximum range for which a radar can detect a target of a given size. The standard definition is max detection range against a target that has a 1 m^2 or 0 dBm^2 surface area reflecting in the direction of the radar. However, radar manufacturers will use different target sizes or, in some cases, do not specify the target size at all.
- *Effective radiated power (ERP)* is the maximum output power that the radar can focus in a specific direction, measured in watts or decibel watts. ERP and maximum detection range are highly correlated. ERP is also an important consideration when attempting to avoid unintentional interference against other nearby EM systems.[4]
- *Field of regard (FOR)* is the horizontal and vertical extent of the radar coverage, denoted in degrees azimuth (horizontal) by elevation (vertical). Radars that rotate 360° have a 360° FOR in azimuth. Some radars do not rotate and instead steer their beams electronically. These are called phased-array radars, and, when stationary, they have less than 180° coverage in azimuth.[5] When providing coverage for a defended asset, FOR plays an important role in determining the number of radars needed and their locations.
- *Beamwidth* is the angular size or width of the radar beam in both azimuth and elevation. Smaller beamwidths improve angular resolution and target tracking accuracy.[6]
- *Revisit rate* is the time required for the radar to scan its entire FOR. For rotating radars, it is the time needed to complete one rotation. Fast-moving targets require faster revisit rates to track or react to targets near the defended area or asset. A phased array can electronically scan its FOR very quickly, often in fractions of a second.[7]
- *Bandwidth:* Radars do not operate at just one frequency but, rather, within a band of frequencies whose span or bandwidth is measured in hertz. Bandwidth has many important implications, including radar range resolution.[8]
- *Range resolution* is the size of each radar cell in the range direction. It is also the minimum range between two targets along the same bearing from the radar for

---

[4] The degree at which a radar can focus its energy is called *antenna gain*; radar ERP equals the antenna gain times the peak radiated power.

[5] Stationary or nonrotating phased-array radars will often cover ~120° in azimuth (60° to the right and left of boresight; see Wolf, undated) and variable coverage in elevation, depending on the size and orientation of the array.

[6] Gain and beamwidth are related by the following approximation: $G \approx 29{,}000 / (\theta \times \varphi)$, where θ and φ are the beamwidths in azimuth and elevation is measured in degrees.

[7] Stationary phased arrays are well suited for near-continuous tracking of targets within their FORs. However, they suffer from reduced FORs when compared with rotating radars.

[8] For most radars, the range resolution (ΔR) equals c / (2 $\times B$), where c is the speed of light (~300,000,000 meters per second) and B is the radar bandwidth in Hz.

which the radar can still resolve them as two individual targets. Range resolution is an important factor when facing multiple UASs operating close together or when attempting to detect or track a target near clutter (e.g., ground, trees, buildings) along the same bearing.

- *Minimum detectable velocity (MDV)* pertains to Doppler and moving-target indicator radars that rely on the target's radial velocity to detect it. Radars with MDVs significantly greater than zero will have difficulty detecting slow-moving or stationary UASs, especially when near large, stationary features.
- *False alarm rate (FAR)* is the rate (number per minute) at which a radar mistakenly indicates the presence of a target also known as a false alarm. False alarms can occur from within the radar itself and from external sources, such as background clutter, birds, or other emissions.[9] A high FAR can drain radar, operator, and other responding elements' resources and degrade performance against real targets.

UAS Target Set Characteristics

We considered three broad classes of sUASs: (1) octo- and quadcopters that can hover or move at very slow speeds, (2) slow-moving fixed-wing gliders, and (3) fast-moving fixed-wing UASs with jet propulsion. Radar detecting and tracking of sUASs will depend to a great degree on the strength of the total sUAS signal return as well as the return from specific components (e.g., rotors) and the environment that the sUAS is operating in. Classification of a target track (e.g., hostile versus non-hostile) is more reliant on specific knowledge of threats, including their radar signatures and patterns, behaviors, and flight profiles. This requires a threat database that includes relevant characteristics that must be continuously updated as new threats emerge.

The strength of a target's signal return depends on its range from the radar and the percentage of incident EM energy that it reflects back toward the emitting radar. The latter is referred to as the target RCS, measured in m^2 or dBm^2. One often thinks of target RCS as equivalent to its physical size or surface area. While there is some correlation, other characteristics—such as composite materials and object shape—can have a significant impact on RCS. Case in point, the B-2 stealth bomber has an RCS that is significantly less than its actual size. Unfortunately, target RCS cannot be represented as a single value because it often varies considerably with geometry (aspect angle between radar and target) and radar frequency. This is because complex targets have multiple scatter points, and the radar waves reflected from these scatter points can interact constructively (in phase) or destructively (out of phase), depending on their relative geometry and the radar wavelength. Figure 5.1 shows how the DJI Phantom 2 Vision RCS can vary with aspect angle.

[9] Radars generate thermal noise within the radar receiver. At times, this noise can be strong enough to exceed the detection threshold, thereby indicating the presence of a target.

Figure 5.1
DJI Phantom 2 Radar Cross-Section (X-Band) as a Function of Azimuth at 0° Elevation

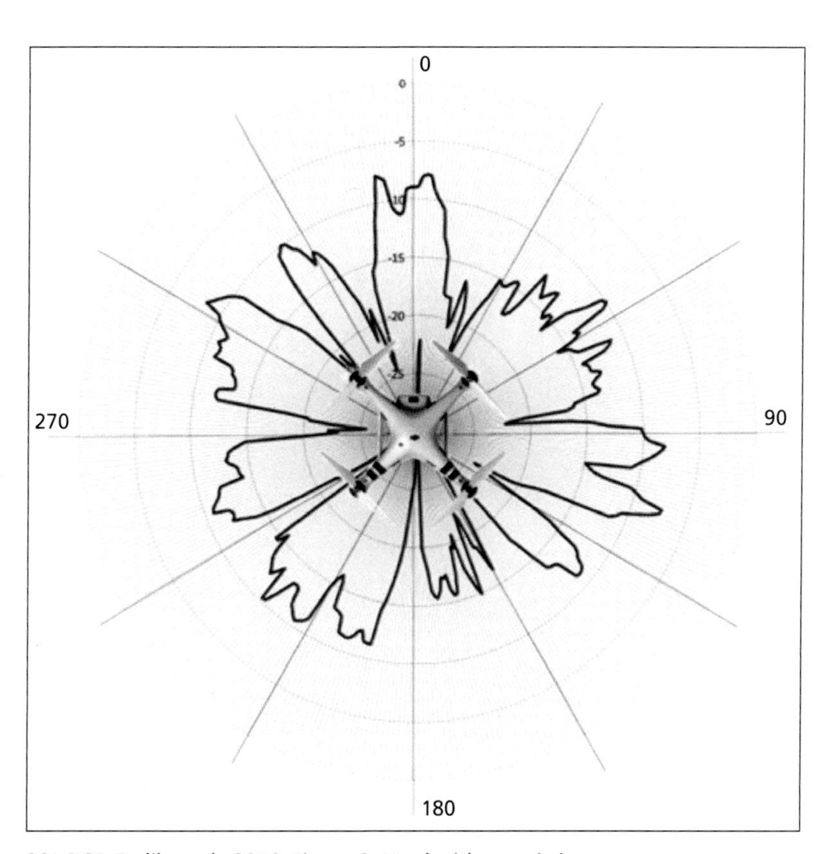

SOURCE: Farlik et al., 2016, Figure 3. Used with permission.

In this case, the radar frequency is 9.3–10 GHz (X-Band) and is looking horizontally at the UAS with no elevation angle. Note that at 0° aspect angle (head-on to the UAS), the RCS is approximately –10 dBm^2 or 0.1 m^2. But at other aspect angles, such as 240°, the RCS decreases by several orders below –25 dBm^2 or 0.003 m^2.

Distant targets or targets moving very slowly (or hovering) may at times sit in one of the low-RCS aspect angles relative to the radar. One solution is to employ multiple radars at different locations with overlapping coverage in the hope that at least one of the radars has an aspect angle where the target RCS is relatively large. However, this may be too costly. An alternative solution is to employ a single radar with a wider frequency bandwidth. Changing frequencies over a wide band also has the effect of changing the shape of the RCS–aspect angle plot. Figure 5.2 shows the effect of increasing the radar bandwidth against a fictional target made up of four cylinders acting as individual scatter points.

Figure 5.2
Comparing Narrowband and Wideband RCS Patterns of a Four-Scatter Target

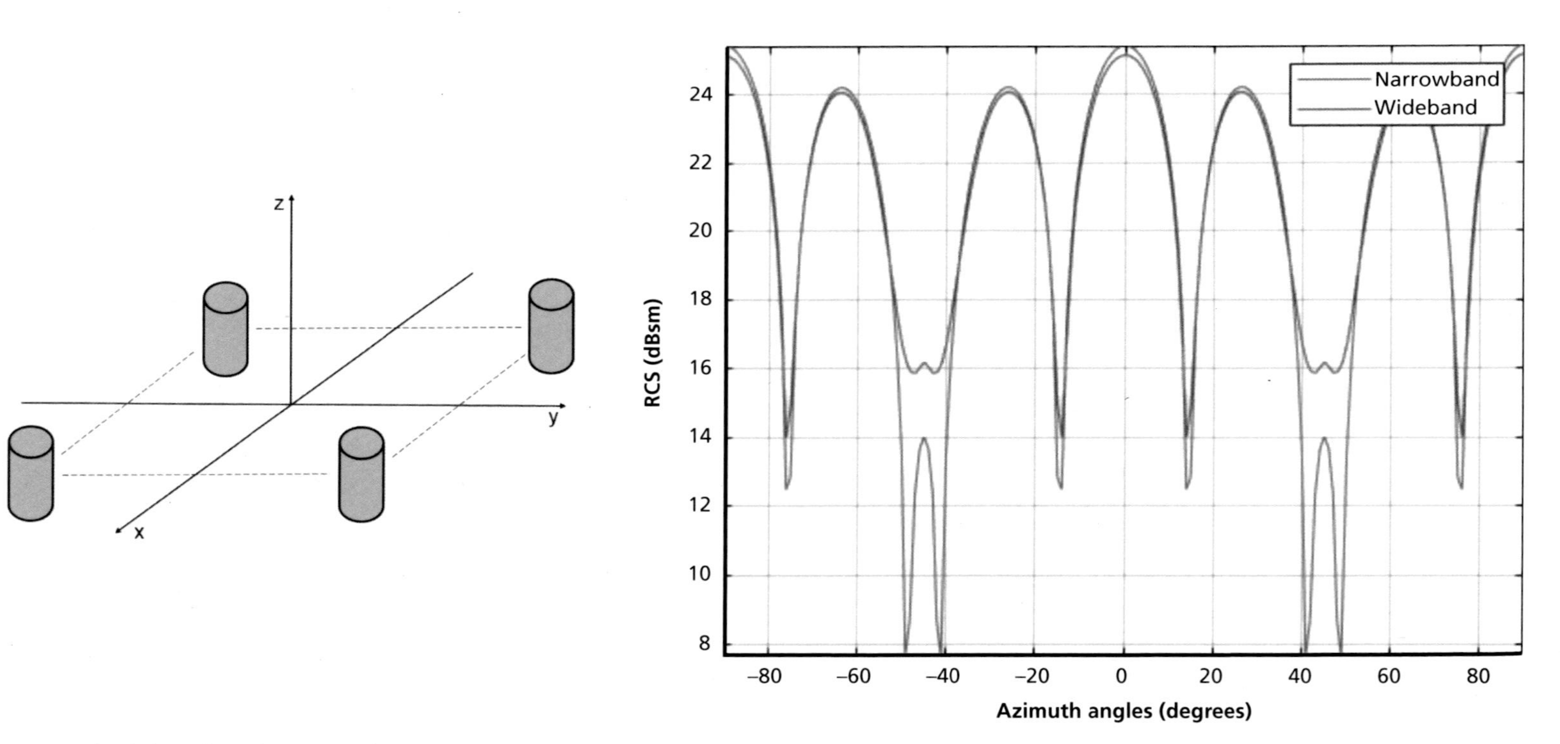

SOURCE: (Right) MathWorks, undated. Used with permission.
NOTE: The graph on the right shows the RCS pattern at 0° elevation for the extended target with the four scatterers.

The narrowband operates at frequency of 300 MHz, whereas the wideband combines frequencies from 270 to 330 MHz with a bandwidth of 60 MHz. Note that for the narrowband, the RCS can vary by as much as 17 dB or a magnitude of 50 times. The wideband varies far less with low RCS values that are 8 dB or six times higher than those for the narrowband.

Atmospheric Effects

Water vapor and oxygen in the atmosphere attenuate and absorb radar energy as it passes thereby reducing radar performance. This effect even occurs in clear weather because of relative humidity. Figure 5.3 shows the extent of atmospheric attenuation from both water vapor and oxygen during clear weather (e.g., no rain) as a function of radar frequency.

Attenuation losses are measured in the decrease of signal strength per kilometer of propagation (dB/km).[10] The attenuation effect generally increases as the frequency increases and is more pronounced at higher frequencies especially above 12 GHz. However, there are notable exceptions below 100 GHz—namely at ~23 GHz and ~52 GHz, at which point peaks in water vapor and oxygen attenuation occur.

Figure 5.3
Atmospheric Attenuation from Water Vapor and Oxygen in Clear Weather

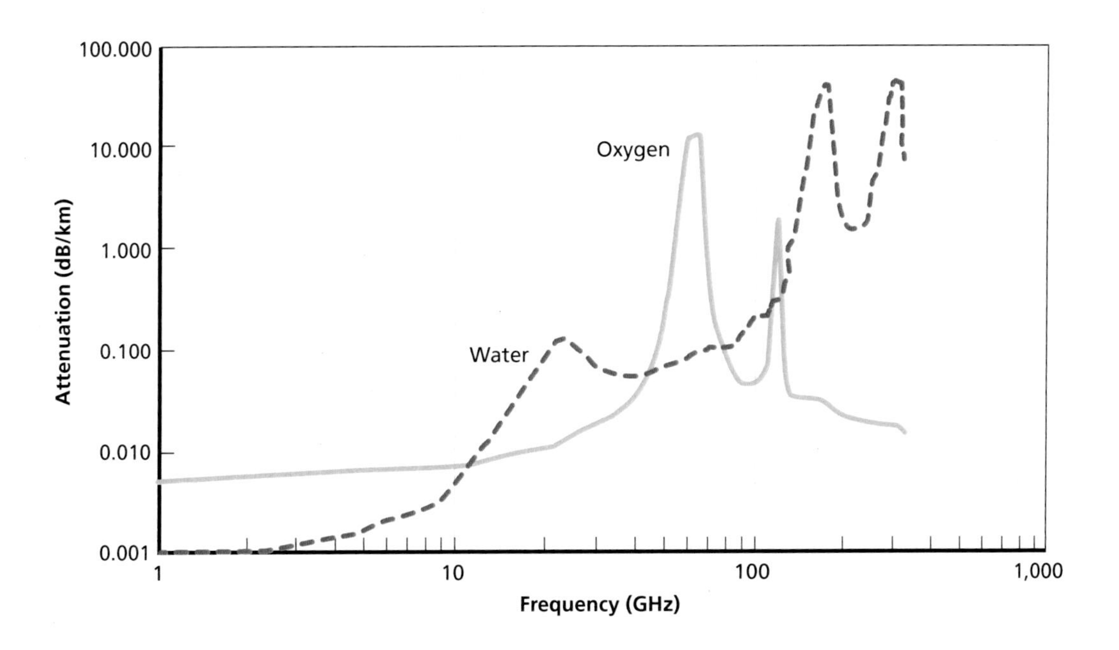

[10] For example, an attenuation loss of 0.1 dB/km implies that the radar energy decreases by 2.3 percent during each kilometer of propagation to and from the target.

Figure 5.4 shows the decrease in power density at ranges out to 40 km from the radar just from atmospheric attenuation (one-way) for S-, X-, Ku-, and Ka-Band frequencies in different environmental conditions: clear (no rain), light rain (2 mm/hour), moderate rain (4 mm/hour), and heavy rain (16 mm/hour). [11]

Clear conditions only include attenuation contributions from water vapor and oxygen, whereas rainy conditions also include the effects from rain drops.[12]

Increasing the radar operating frequency (decreasing the radar wavelength) allows it to focus more energy into a tighter beam, thereby increasing the radar ERP and, in theory (assuming that both the radiated power and aperture size are held constant), increasing the detection range as well.[13] However, as shown in Figure 5.4, higher frequencies generally entail more atmospheric attenuation, which decreases the energy that reaches the target (and back to the radar), thereby decreasing the radar detection

Figure 5.4
One-Way Radar Power Density Degradation from Atmospheric and Rain Attenuation

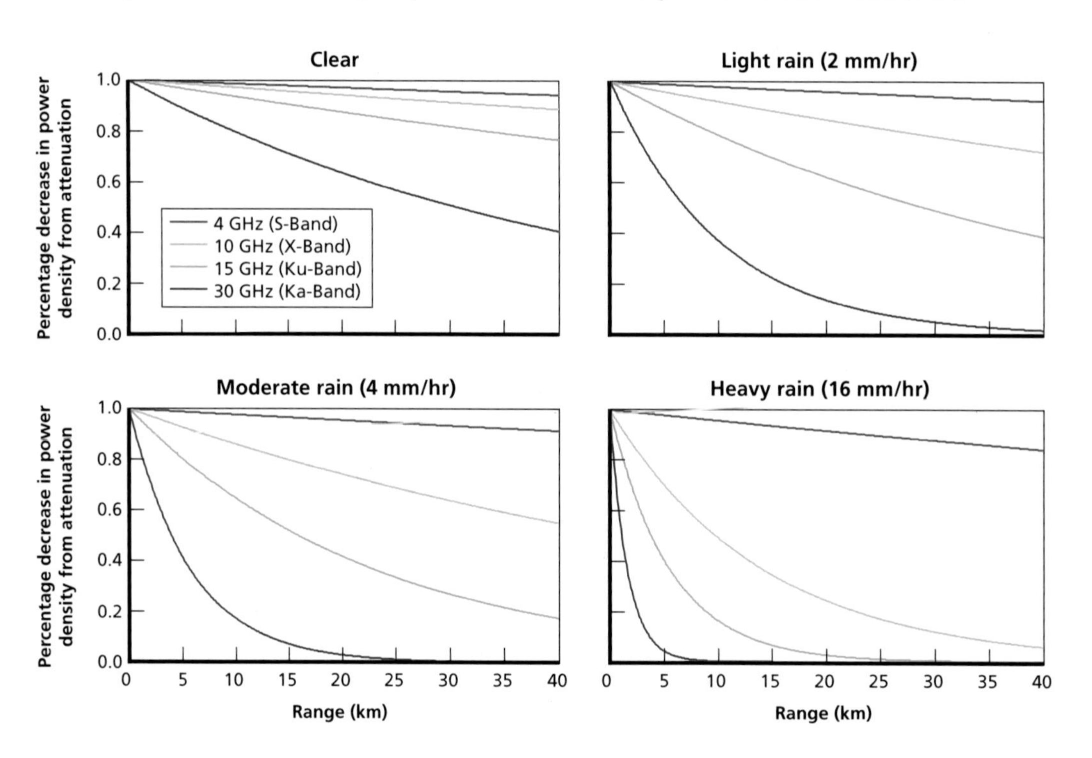

---

[11] Power density also decreases with range as the wave front spreads out spherically. If R is the range from the radar antenna in meters, then the power density (watts per m^2) decreases on the order of $1 / R^2$ from the spreading effect. Figure 5.4 shows power loss from atmospheric attenuation only and not from spreading.

[12] For a discussion on rain-induced attenuation, see International Telecommunications Union, 2005.

[13] Specifically, antenna gain (G) equals $4 \times \pi \times A_e / \lambda^2$, whereas the effective size of the antenna aperture in m^2 and λ is the radar wavelength in meters.

range. So, there is an interesting trade-off. Most military-grade target-tracking radars operate in X-Band (8–12 GHz) where the antenna gain is high enough for good angular resolution, and, at the same time, the attenuation is weak enough to allow for target tracking at ranges of hundreds of kilometers. But against the sUAS threat, candidate radars do not have to detect and track at such long ranges, making higher-frequency bands (e.g., Ku, Ka) more attractive. Moreover, the increased attenuation may actually be beneficial by limiting the interference region against other EM receivers. On the other hand, poor weather conditions severely diminish radar performance when operating at the higher-frequency bands.

Example Radars for C-UAS

We provide performance data for three prospective C-UAS radars that vary considerably in size and, presumably, effective radiated power: DeTect Harrier/DSR-200, Gryphon R1400, and Fortem Technologies DAA-R20. Table 5.1 contains a partial list of their relevant parameters. Some data were not available for this analysis.

Based on the data in Table 5.1, we can only estimate the Gryphon R1400 radar's performance with an additional assumption that SRC's definition of *sUAS* implies

Table 5.1
Performance Parameters for DeTech Harrier/DSR-200, SRC Gryphon R1400, and Fortem Technologies DAA-R20 Radars

Parameter	DeTech Harrier/ DSR-2000	SRC Gryphon R1400	Fortem Technologies DAA-R20
Detection range (km)	26–48[a] 7–11 4–6	< 27[b] <10	1.5[c]
FOR: azimuth (degrees)	360	?[d] 360	120
FOR: elevation (degrees)	—	—	40
Scan rate (seconds)	—	0.5–3[d] 2–6	—
ERP (dbW)	—	56	—
Peak power	—	512	—
Frequency band	S or X	X	—
Bandwidth (MHz)	—	5	—

SOURCES: Publicly available manufacturer data.

[a] Stated detection ranges against large, medium, and micro-UASs.

[b] Maximum detection ranges against small manned aircraft and sUASs.

[c] Stated detection range against 1.0 m^2 (0 dBm^2) target.

[d] The literature implies that the radar can remain stationary or it can rotate. The scan rates are for stationary and rotating, respectively. FOR while stationary was not available.

a UAS with an average RCS of 0.1 m^2 (–10 dBm^2).[14] Figure 5.5 shows the Gryphon R1400's detection ranges against targets ranging in size from 0.01 m^2 (–20 dBm^2) to 100 m^2 (20 dBm^2) under different environmental conditions: clear (no rain), light rain (2 mm/hour), moderate rain (4 mm/hour), and heavy rain (16 mm/hour).

During clear weather, the R1400 can detect larger targets (>10 dBm^2) at ranges beyond 30 km and can detect very small targets (–20 dBm^2) at approximately 5 km. Light and moderate rain has some effect on detection ranges, especially against larger targets. However, heavy rain has a more profound effect, decreasing detection ranges by as much as 50 percent for large RCS (20 dBm^2), 30 percent for medium RCS (0 dBm^2), and 15 percent for small RCS (–20 dBm^2). Based on its parameters, we estimate that the R1400 has horizontal and vertical beamwidths of 5.4 and 8.1 degrees and a range resolution of 30 meters. Note that these calculations assume that no background clutter is near the target and that the FAR is constant and relatively low across all conditions. Often, however, weather effects such as rain substantially increase the FAR. Moreover, the nuisance false alarms (e.g., birds) can increase the FAR even during clear weather. To reduce the FAR in the presence of rain, nuisance false alarms, and background clutter, radar detection thresholds rise, thereby reducing detection ranges against real targets from those shown in Figure 5.5, especially in rainy conditions.

Figure 5.5
Estimated Detection Ranges for the Gryphon R1400 Radar Under Varying Environmental Conditions

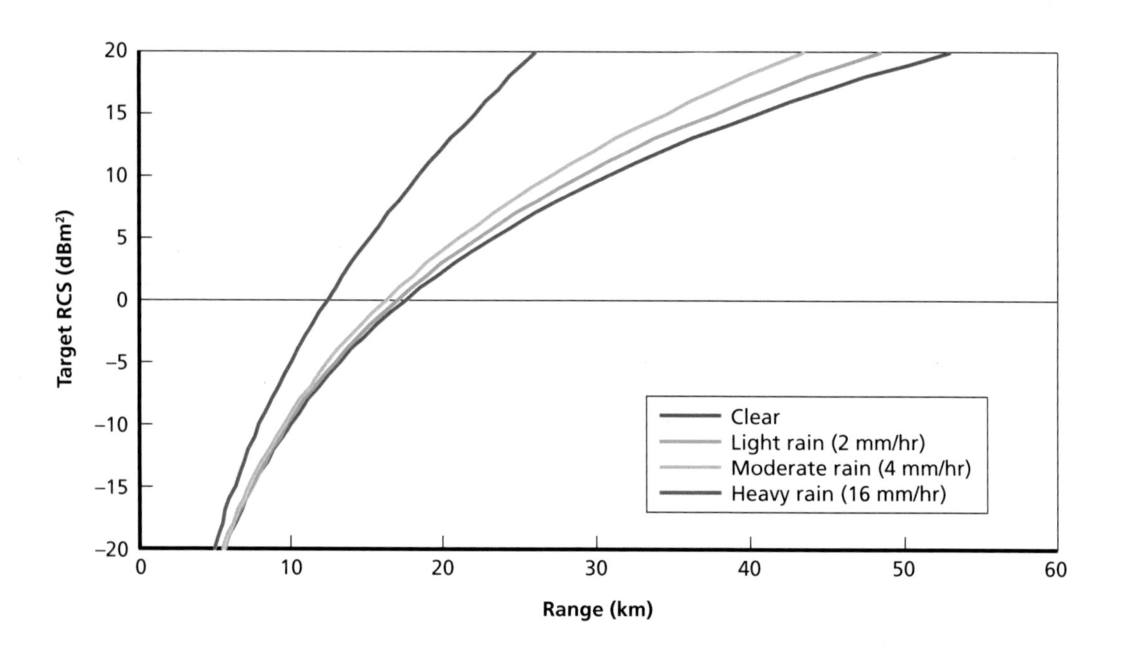

[14] For detailed performance analysis, we would need additional data, such as the probability of detection at a given range, the FAR, and the signal-to-noise ratio threshold.

Doppler radars can extract moving target signatures from background clutter by detecting the Doppler frequency shift in the return signal. This can be from the movement of the sUAS itself or from individual components, such as rotors, depending on their size or RCS. For example, a nine-inch aluminum rotor blade of a DJI Phantom Vision 2+ has an RCS of –20 to –8 dBm^2 when its orientation is perpendicular to the radar line of sight.[15] Small UASs, especially those with moving components, generate unique Doppler signatures that can be used to help classify potential threats based on their radar return.

Future Threat Improvements and Countermeasures Versus Radar

Passive and active measures are potentially available to threat actors that degrade radar performance against sUASs. As happened with manned aircraft, sUAS manufacturers may discover new designs that reduce RCS across a larger proportion of aspect angles. We saw in Figure 5.5 that 20 dB reductions in RCS can reduce detection ranges by 50 percent or more. Furthermore, materials such as plastic tend to absorb EM energy in the RF range rather than absorb it. Experimentation can lead to new UAS shapes, designs, and composite materials that decrease vehicle RCS, as happened with the single-wing or "batwing" aircraft, such as the B-2. Low RCS coupled with high velocity could reduce reaction times against threat UAS, thereby increasing their probability of success; however, this comes with important performance trade-offs.

Improved navigation and maneuverability may assist future UASs in utilizing terrain features, such as wooded areas or "urban canyons," to mask their presence. Doppler processing can aid in detecting targets near large features, such as trees or buildings. However, targets that hover or move at speeds below the radar MDV may still be still be hidden within the surrounding clutter unless the sUAS has faster-moving components (e.g., rotors) that can be detected by the radar. But, in some cases, manufacturers are placing ducts around rotors to limit noise generation, making rotor detection more difficult. Finally, DRFM is a relatively new electronic warfare (EW) technique in which an EW receiver digitally records a pulse received from an enemy radar and then retransmits the pulse in an effort to deceive or spoof the enemy radar, for example, by creating false targets or through active cancellation to blank out real targets. This capability is currently available on military systems only. However, if it *does* enter the commercial market in the next ten years, it could further complicate radar detection and tracking of threat UASs.

Counters to these future threats include radars with wider bandwidths, advanced digital signal processing, and frequency and waveform agility. Figure 5.2 illustrated how wideband waveforms can mitigate RCS nulls due to target shape and geometry better than narrowband waveforms. Wideband waveforms also have better (smaller) range resolution. Not only does this enhance target tracking accuracy, but it also

---

[15] RCS values range from –40 to –20 dBm^2 across other orientations (Ritchie et al., 2016).

reduces the size of the radar cell (in range), decreasing the chance that a target and large clutter echo are within the same cell. Unfortunately, increasing the radar bandwidth would reduce the radar pulse width and energy.

Digital signal processing has vastly improved the ability to detect targets in the presence of clutter. One technique maps out the clutter within the radar FOR and stores it digitally. Slight changes in the clutter return from a particular cell may indicate that a target is present, similar to how moving-target indicator technology works. Furthermore, digitally storing the multiple radar returns and then correlating them using big data techniques can further facilitate the process of deciphering real targets from false alarms and clutter. Finally, pulse-to-pulse frequency and waveform agility will make it difficult for DRFM systems to anticipate and mimic an inbound radar pulse in advance.

EO/IR Sensors

EO/IR sensors also operate in the EM domain but at much higher frequencies than radar where the wavelengths are measured in microns or millionths of meters (denoted by μm, e.g., 10^{-6} meters). Specific bands include the following:

- visible (EO): 0.38 μm to 0.75 μm
- short-wave IR (SWIR): 1.0 μm to 3.0 μm
- midwave IR (MWIR): 3.0 μm to 5.0 μm
- Long-wave IR (LWIR): 8.0 μm to 14.0 μm.[16]

In this section, we focus on passive EO/IR sensors that detect illumination reflected off targets from external sources, as well as energy emitted from the targets themselves. There are two broad classes of EO/IR sensors—imaging and non-imaging. Imaging sensors have enough resolution to characterize the internal detail of objects within some range of the sensor. They can create still-frame or full-motion video images that can be used to detect, classify, identify, and track targets, depending on the resolution, range, and other factors. Non-imaging sensors treat targets as point targets that reflect and emanate energy without resolving internal details of the target. As such, non-imaging sensors can detect and track potential targets, but generally do not have the resolution to classify or identify them. On the other hand, nonimagers tend to have much wider fields of view (FOV) and can search the sky for targets much more efficiently than their imaging counterparts. IR search and track (IRST) systems are examples of non-imaging sensors. Most modern imaging systems are digital

[16] These four bands range from 21 THz (LWIR) to 790 THz (EO) in frequency where one THz equals 1,000 GHz.

where each still-frame image is broken down into individual pixels with digital values assigned to them based on the image return.

UAS Target Set Characteristics

Physical size of the target directly impacts EO/IR sensor performance for both imaging and non-imaging sensors. Target extent (as viewed from an imaging sensor) along with the range from sensor to target determines the number of pixels or detector elements receiving photon energy from the target. As the number of pixels on target increases, so too does the probabilities of target detection, classification, and identification. Radiant intensity also increases with target size—specifically, the target's surface area. Both quantities will vary with target aspect angle relative to the sensor. Contrast and differences in radiant intensity between the target and surrounding background are equally important in determining sensor performance. Backgrounds with similar signatures as the target will tend to mask them from the sensor. Finally, battery-powered vehicles such as octo- and quadcopters and fixed-wing gliders have much lower radiant intensities than their jet-powered counterparts and will be more difficult to decipher from the environment via non-imaging IRST systems.

Atmospheric Effects

Like radar, atmospheric elements (water vapor, oxygen, and other molecules and particles) degrade EO/IR sensor performance refracting, absorbing, and scattering photon energy. Figure 5.6 shows the transmittance across the four EO/IR bands. Although they present the same effect, Figure 5.6 differs from Figure 5.3 in that the former depicts the percentage of unattenuated energy per kilometer of propagation (left axis) instead of attenuated or lost energy per kilometer (in dB/km). For comparison purposes, we have added an inverted vertical axis on the right side showing the equivalent energy losses in dB/km. The lowest attenuation occurs at about 4 microns with approximately 90-percent transmittance per kilometer, or an energy loss of 0.5 dB/km, which is significantly more attenuation than at radar frequencies less than 50 GHz (see Figure 5.3). But, unlike radar, passive EO/IR sensors only require one-way energy transmission, which partially mitigates the higher attenuation rates.

Rain and fog increase attenuation and further reduce detection, classification, and identification ranges. Since rain drops are typically much larger than EO/IR wavelengths, the additional attenuation causes by rain is virtually the same across the EO and IR bands. For example, light rain (2 mm/hour) causes an additional 30 percent loss in transmittance per kilometer (1.5 dB/km); moderate rain (4 mm/hour) leads to an additional 45 percent loss (2.5 dB/km); and heavy rain (16 mm/hour) causes an additional 80 percent loss (7 dB/km). Unlike rain, fog and haze comprise particles that are comparable in size to the wavelengths of visible light. So, the effects, which depend on the severity (measured in terms of visibility), vary with wavelength. For EO, fog and haze increase attenuation losses by 6–300 dB/km as the visibility ranges from five

Figure 5.6
Atmospheric Attenuation for EO/IR Bands During Clear Weather

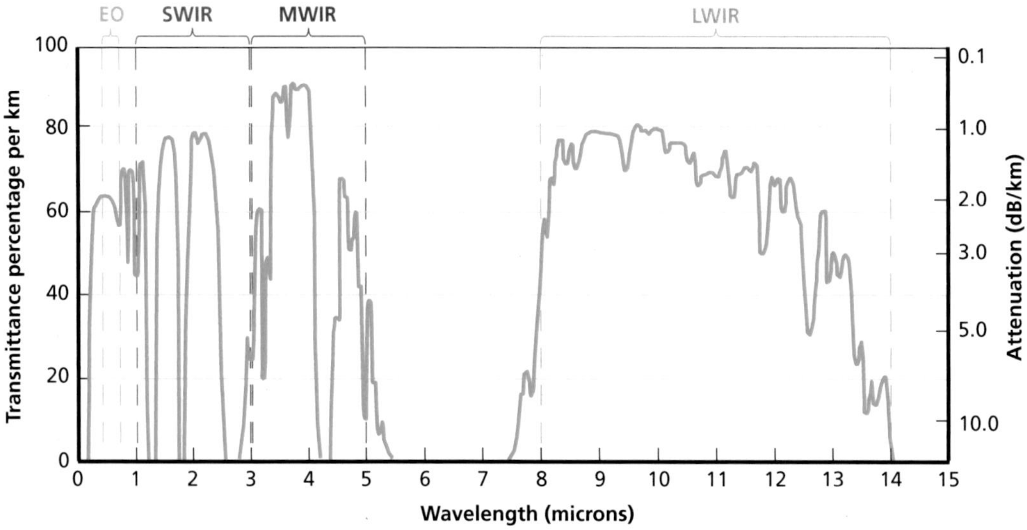

SOURCE: Chen, 1975.

kilometers to just 60 meters. For LWIR, losses range from 0.25 to 150 dB/km (Chen, 1975).

Imaging Sensor Performance

Imaging sensors (EO and IR) have enough resolution to determine internal features of objects if the objects are close enough to the imaging sensor. Angular resolution (horizontal and vertical) is a function of the pixel pitch (size of the individual detector elements) and the lens focal length.[17] To determine the imaging resolution (in meters × meters) at some range, multiply the range (meters) by the angular resolution (in radians). For example, a camera with angular resolutions of 0.115 degrees in the horizontal and vertical will create a two-by-two meter resolution cell at a range of one kilometer, or 1,000 meters. And objects or object features that are smaller than two meters in both dimensions may not be resolvable at that range. Both the angular resolution and angular field of view are inversely related to the lens focal length.[18] So, for the same-sized detector array and pixel pitch or count, one can only improve the resolution while decreasing the sensor field of view. Consequently, most low-cost, high-resolution EO/

---

[17] Angular resolution (horizontal or vertical), denoted *IFOV*, equals $2 \times \mathrm{Tan}^{-1}(d / 2 \times f)$, where d is the pixel pitch and f is the lens focal length.

[18] Angular field of view (horizontal or vertical), denoted *FOV*, equals $2 \times \mathrm{Tan}^{-1}(D / 2 \times f)$, where D is the size of the detector array (horizontal or vertical) and f is the lens focal length.

IR imagers with sufficient range have relatively small fields of view. As such, they will likely be most effective when cued by other sensors.

According to Johnson's Criteria, object detection, recognition, and identification within an image depend on the average number of resolution cells or pixels that can be placed on the object in both spatial dimensions. Specifically, under ideal conditions, 50-percent probability of detection, recognition, and identification is as follows:[19]

- Detection (an object is present) requires at least *three* pixels on the object in both dimensions.
- Recognition (the type object can be discerned, e.g., person versus car) requires at least *ten* pixels on the object in both dimensions.
- Identification (a specific object can be discerned, e.g., woman versus man, pickup truck versus sedan) requires at least *16* pixels on the object in both dimensions (see Koretsky, Nicoll, and Taylor, 2013).

Figure 5.7 shows the detection, recognition, and identification ranges against a DJI Phantom based on these criteria as a function of the sensor focal length-to-pixel pitch ratio (mm/µm).[20] For example, an imaging sensor with a focal length and pixel pitch of 100 mm and 25 µm (ratio of four) will detect, recognize, and identify the DJI Phantom 4 at ranges of 257, 80, and 48 meters with 50-percent probability. In comparison, 200-mm focal length and 10-µm pixel pitch (ratio of 20) lead to equivalent performance ranges of 1,290, 400, and 240 meters.

It should be noted that Johnson's Criteria are applicable only under ideal conditions and do not account for other factors that can significantly degrade imaging performance from what is shown in Figure 5.7. Such factors include contrast (EO) and temperature difference (IR) between the target and the immediate background. Separating targets from backgrounds with very similar signatures depends on the system sensitivity (EO) and minimum resolvable temperature difference (IR). Also, weather conditions such as rain and fog reduce both the target and background signals that reach the sensor, which has the net effect of further reducing the contrast and temperature difference between them. On the other hand, advanced post-image processing techniques are available that can sharpen an image in real time, increasing the image contrast and making it more interpretable to the human eye. Finally, there are image-based tracking algorithms that can automatically adjust the sensor gimbal to keep the

[19] Johnson's Criteria are described as the number of line pairs that can be placed on the target along its smallest dimension. Under ideal conditions, the Nyquist Sampling Theorem indicates that a minimum of two pixels is required to successfully capture a line pair. We have converted line pairs to pixels assuming ideal conditions. Less-than-ideal conditions would require more pixels on target than stated here.

[20] DJI Phantom 4 is $25 \times 25 \times 19.3$ cm in size. The calculations in Figure 5.7 assume that the sUAS is flying directly toward the sensor.

Figure 5.7
Imaging Sensor Performance Against the DJI Phantom 4 Under Ideal Conditions

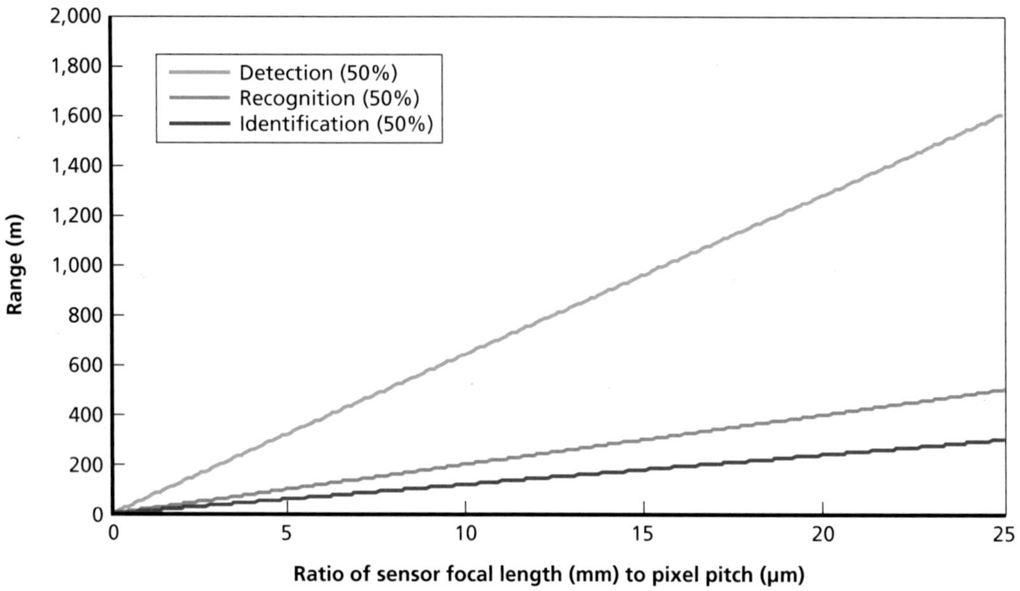

target in the center of the field of view after initially being designated by a human operator.

Non-Imaging Sensor Performance

Non-imaging sensors detect and possibly track objects as point targets without necessarily resolving any internal detail of the targets. Examples include IRST systems like those employed on fighter aircraft that detect and track other fighter aircraft, and the U.S. Space-Based Infrared System (SBIRS) that detects and tracks ballistic missile launches.

IRST systems can detect aircraft at long ranges because they tend to be much hotter (particularly the engines) than the background sky. Their effectiveness against sUAS will depend on their sensitivity as well as the UAS heat contrast against the background sky. Researchers from the University of Zagreb recently showed that even inexpensive thermal systems (which are still capable of sub–degree centigrade differentiation) can be used to detect sUASs in many conditions (Andraši et al., 2017). These experiments determined batteries to be the greatest source of heat, as they received less airflow cooling than motors (it is unclear how much drag at high speeds is a heat source). Commercial users could see improved battery life from reduced battery temperatures, so there is some incentive for manufacturers to pursue this, and it could lead to reduced thermal signatures in the future.

Most non-imaging sensors operate in the MWIR and LWIR bands where most of the energy is emitted from the targets themselves. The energy radiated or radiance (measured in watts per square centimeter of target surface area per steradian, e.g., $W/cm^2/ster$) depends on the target temperature and emissivity as well as the energy wavelength.[21] For objects near room temperature (300 K), the radiance is strongest within the LWIR band. Detecting and then tracking targets of interest depend on differences in radiance signatures between the target and the surrounding background as well as the thermal sensitivity of the sensor. Non-imaging systems, such as IRST systems, can employ complex algorithms to automatically detect and track potential targets without a human-in-the-loop. These algorithms utilize detection thresholds (specific power delta between target and background radiance intensities) that yield both a probability of detection (for a given delta) and a false alarm rate, depending on the signal-to-noise ratio within the sensor.

The standard equation for determining target detection range is given in Equation 5.1:

$$SNR = \frac{ENSQ \times A_e \times 10^{-0.1\times 0.001\times \beta \times R} \times 0.01 \times D^* \times \Delta I}{10^{-6} \times R^2 \times \sqrt{A_d \times \Delta f}}, \quad (5.1)$$
$$= \Gamma \frac{10^{-0.1\times \beta \times R} \times \Delta I}{R^2}$$

where one solves for the detection range, R, for a given signal-to-noise threshold (see Nicholas et al., 2018). The variables (units) in the equation above are defined as follows:

1. *SNR:* signal-to-noise ratio (unitless)
2. *ENSQ:* also known as ensquared energy, which is the percentage of total power from a distant point source that is incident to a single pixel, assumed to equal 0.9671
3. A_e: effective aperture size of the optic lens, or the physical aperture size (m^2) multiplied by a factor (~0.8) that accounts for energy lost due to reflection and other factors
4. β: attenuation losses (dB/km)
5. D^*: sensor detectivity measured in Jones × ster
6. ΔI: difference or delta between target and background radiant intensities (W/ster)
7. R: range from target to sensor (m)
8. A_d: area of the individual detector or pixel (μm^2)
9. Γ: consolidation of sensor-related variables into one variable for computational purposes.

[21] A steradian is a unit of angular measurement in 3D space equivalent to a radian in two-dimensional space.

Nicholas et al. (2018) developed a model based on Equation 5.1 to predict sensor performance against sUASs under a variety of background conditions. The sensor under consideration operated in a portion of the LWIR band (8–10.5 μm) with a Γ-value of 6.08×10^9 and a 30° field of view.[22] A Sky Viper sUAS served as the target. The researchers analyzed four background conditions: cold sky, warm sky, cloudy sky, and tree cover. The researchers determined the following differences in radiant intensities between the target and the four backgrounds (ΔI):

- cold sky: 277 mW/ster
- warm sky: 310 mW/ster
- cloudy sky: 91 mW/ster
- tree cover: 52 mW/ster.

They also assumed that detection required an SNR of at least eight. Using Equation 5.1, Figure 5.8 plots detection range as a function of ΔI for the sensor (with Γ equal to 6.08×10^9) and SNR equal to eight under clear conditions, moderate rain (4 mm/hr), and fog (300-meter visibility). These values assume no obstructions between the sensor and target.

Figure 5.8
Detection Ranges Versus Sky Viper UAS Under Clear Weather, Moderate Rain (4 mm/hour), and Fog with 300-Meter Visibility

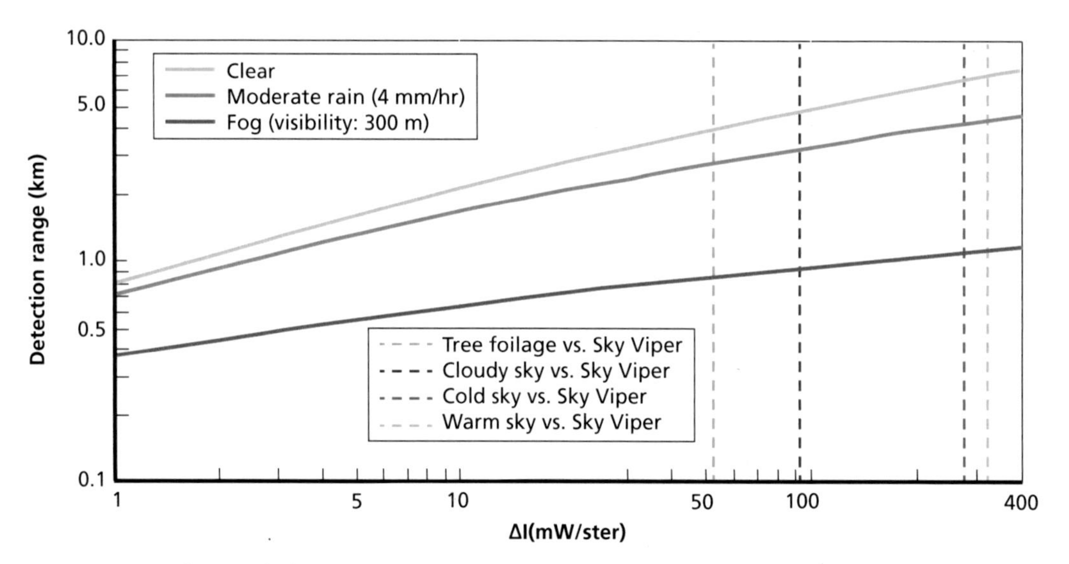

SOURCES: Data from Nicholas et al., 2018; Chen, 1975.

[22] For more details regarding the specific sensor parameters, see Nicholas et al., 2018.

Under clear atmospheric conditions, detection ranges varied from 3.9 km (tree foliage) to 6.9 km (warm sky). Detection ranges during moderate rain went from 2.8 km to 4.3 km across the four backgrounds, whereas detection ranges during fog (300-meter visibility) fell to 0.9 km to 1.1 km. It should be noted that Sky Viper is about 19.5 × 13 × 3 inches in size and weighs only 2.9 pounds. Larger sUASs may produce larger radiant intensity differences with the various backgrounds and, in turn, increase the detection ranges from those shown in Figure 5.8. Finally, sensors with higher Γ-values have longer detection ranges, but one must also consider other parameters, such as sensor field of view, when countering an sUAS threat.

Threat Countermeasures Versus EO/IR

Physically smaller UASs are harder to see. If a threat mission can be accomplished with smaller (and perhaps many) UASs, then this could complicate EO/IR detection, tracking, and identification. Also, flying low and utilizing the terrain (e.g., trees, urban canyons) can thwart EO/IR sensors. For many years, the U.S. military has attempted to reduce the visibility of its aircraft by reducing their contrast with the background. It has also strived to reduce their IR signatures. SUAS manufacturers can certainly pursue the same goals. Automatic target detection and tracking (possibly with a human in the loop) and real-time image processing can offset some sUAS improvements. Allowing machines to scan an image and cue humans to possible targets will lessen the burden on the human operator. Recent military systems, like the LITENING and SNIPER advanced targeting pods, have implemented real-time image processing that immediately sharpens the image and increases the contrast between objects and the background and within the object itself, leading to a higher probability of target identification. However, such processing techniques also sharpen the noise in an image, increasing the visual clutter for the operator and thereby potentially reducing identification effectiveness against small or faint targets. It is not clear whether similar processing techniques are available in the commercial market yet. If so (or if these algorithms can be made available to DHS), then this would increase performance against sUASs at longer ranges and in challenging environmental conditions.

ELINT Sensors

Unlike radar, ELINT sensors do not emit; instead, they passively detect signals of interest including those emitted by the UAS and/or its control element. Depending on the sensor suite, ELINT sensors can locate emitters of interest with different degrees of fidelity from just a bearing to a three-dimensional location with good accuracy (< 100 meters from the true location). In a potentially crowded electromagnetic environment, ELINT sensors must scan the frequency spectrum to detect and identify signals of interest. Important performance parameters include sensitivity and bandwidth per scan. It must have the sensitivity to detect weak signals at preferably longer ranges and enough bandwidth to cover the target signal bandwidth. Signal identification depends predominately on *a priori* knowledge of the signal of interest, including its frequency, bandwidth, and modulations. ELINT sensors rely on digital libraries containing waveform parameters of threat emitters. If a signal is detected, it is quickly compared to the signals within the library for identification.

ELINT sensors also suffer from false alarms, including nuisance false alarms from wireless internet hotspots, GoPro camera systems, and other emitters. The primary filtering method is library lookup; if an incoming signal does not match any signal within the signal library, then it is simply ignored by the ELINT sensor. Unfortunately, an unknown signal may also originate from a threat sUAS and just ignoring it may not be an acceptable action. However, reacting to every unknown signal, especially in a dense EM environment, may yield an unwieldy number of false alarms to process and interpret. This stresses the importance of understanding potential threat EM signatures and updating the threat EM libraries and databases as quickly as possible.

One ELINT sensor can obtain a bearing on a threat within several degrees, depending on the size and sensitivity of the ELINT receiver and the threat emitter frequency. However, it cannot provide range-to-target information. Multiple sensors listening to the same signal can obtain coarse range and altitude information via triangulation. The target location error becomes smaller with more sensors and a wider angular spread between the sensors relative to the target. At longer ranges (> 20 km), target location error can be on the order of one to several kilometers. Other techniques such as time difference of arrival (TDOA) can achieve much better target location accuracy. TDOA involves multiple ELINT sensors that compare the arrival times of a threat emitter pulse, which creates a set of isochrones whose intersection is the target location.[23] One needs four ELINT sensors to generate a target coordinate and altitude. And, unlike relying on bearings only, TDOA accuracy is mostly independent of target range.[24] Once again, the TDOA is more accurate when the sensors have wider angular separation. At a 90-degree separation, errors as small as tens of meters can be attained. Such errors are small enough to directly cue high-resolution imagery sensors for tracking and identification.

Threat Countermeasures Versus ELINT

An obvious sUAS countermeasure is for the sUAS and its ground-control segment to simply not emit. For some missions, the sUAS can fly a preplanned route with the

[23] This assumes that the threat emitter transmits its signal in discrete pulses. Some emitters known as continuous wave (CW) emitters emit a signal continuously. TDOA is still possible against CW emitters, but requires more complicated processing that compares time-stamped in-phase/quadrature modulation data from the target emitter.

[24] Except at ranges were the target emitter signal strength drops near or below the sensitivity levels of the ELINT sensors.

aid of GNSS without relying on external communication. Other missions that rely on off-board transmittance, such as video imagery to a ground element, can be easily detected. As state earlier, successful ELINT relies on some knowledge of enemy signals. If threat actors can somehow manipulate their emissions to mimic other nearby nonthreat emissions, this may complicate efforts to distinguish them from the background signal environment. Altering a signal to a completely new waveform may not work because ELINT sensors can be programmed to cue on unrecognized signals. Finally, the threat may employ multiple sUASs, with some acting as emitter decoys to draw attention from the primary sUAS. The emitting sUAS may emit for a purpose (e.g., sending information) or act strictly as a decoy.

Acoustic Sensors

For acoustic signatures, sound power levels decrease by approximately 6 dB for every doubling of distance between source and receiver. This limits acoustic detection to relatively close ranges: a few dozen meters or less for sUASs with electric propulsion, depending on the particular design, the sophistication of the detection system hardware and software, and the amount of competing noise in the environment.[25] The frequency spectrum of the system's noise also plays a role in determining detection range; in the air, lower frequencies are attenuated less than higher frequencies. Again, this depends on the design parameters of the UAS, but as Figure 5.9 shows, sound pressure levels are generally higher in the medium band of the frequency spectrum.[26]

There is a market demand signal for quieter sUASs, since many recreational and professional users would like to fly at events or in serene natural settings, and sUAS manufacturers have started developing (and advertising) noise-reducing features, such as more sophisticated propeller shapes (Figure 5.10), quieter motors (which often also offer higher efficiency), and rotor shrouding.

Using more-sophisticated mission profiles, such as letting a fixed-wing sUAS glide toward its target with the engine switched off or having a rotary-wing sUAS perch on an elevated surface without using its engines, can also reduce the likelihood of acoustic detection. However, acoustic signatures still play an important role in C-UAS planning because they can alert humans to the presence and approximate location of a nearby UAS without the need for additional equipment.

---

[25] Note that especially in the case of a high noise background, detection can be aided by looking for characteristic changes over time (modulation) of the sound pressure signal; however, this requires more sophisticated sensors and processing techniques.

[26] However, UASs also emit time-variable ultrasonic noise (Fu, Kinniry, and Kloepper, 2018) that could be used for short-range detection, characterization, or localization. Another potential approach to acoustic detection is to look for UAS-generated vibrations in the ground or in structures (seismic detection).

Figure 5.9
Quadcopter Sound Pressure Level, by Distance and Altitude

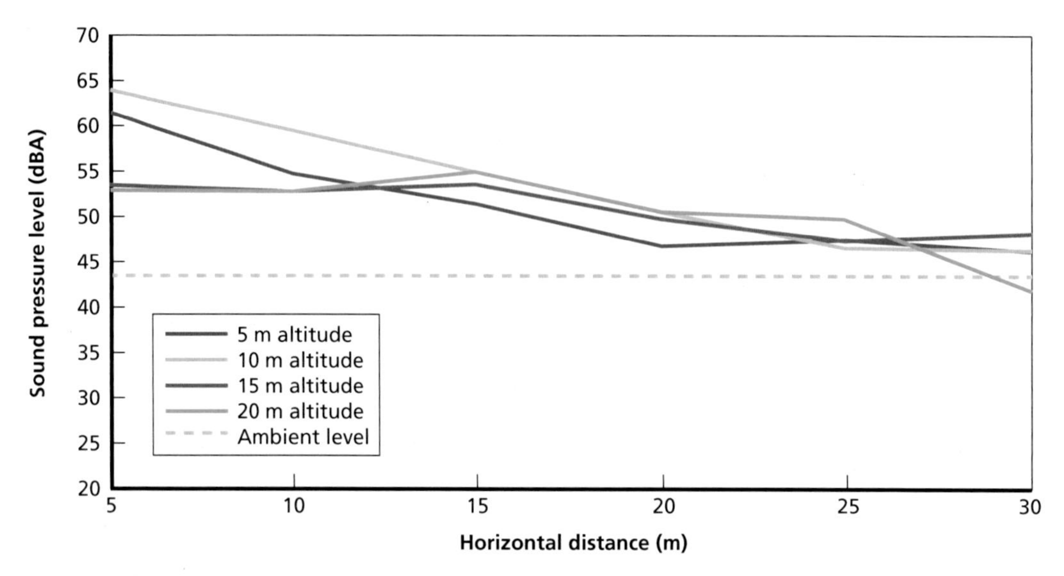

SOURCE: Kloet, Watkins, and Clothier, 2017 (CC BY-NC 3.0).

Figure 5.10
Sound Pressure Level Reduction Achieved by Sophisticated Rotor Blade Shape

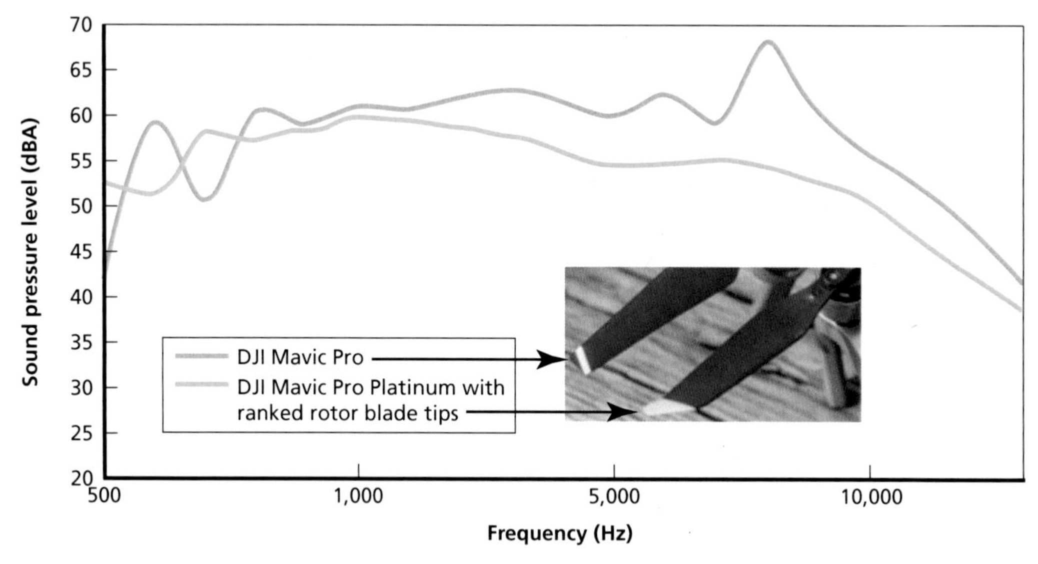

SOURCE: Compiled from DJI promotional materials.

Commercial sUAS manufacturers will continue to reduce sUAS noise emissions due to market pressures and because refining rotor blade shapes or using higher-quality, quieter motors is relatively simple but can lead to significant reductions in noise. For these reasons, one C-UAS system manufacturer that we consulted for this study no longer recommends acoustic sensors as a primary approach to sUAS detection; however, acoustic sensors can still play a role as part of a multimodal sensor system, and other manufacturers still offer them (see, e.g., Squarehead Technology, undated). Array-based acoustic detection systems that consist of a large number of microphones installed both inside and outside a protected zone, comparable to the "ShotSpotter" system for gunfire detection (Shotspotter, undated), are another option based on acoustic signatures.

LIDAR Sensors

LIDAR sensors detect targets by illuminating them with pulsed laser light and collecting the reflected pulses. LIDAR sensors can accurately measure target distances based on the round-trip times of the reflected laser pulses. LIDAR is very similar to radar; both are active sensors that illuminate objects with electro-magnetic energy within their field of view. However, radars operate within a portion of the RF spectrum (less than a 100 GHz) whereas LIDAR sensors typically utilize ultraviolet, visible, or near-IR light to illuminate their targets. And unlike most radars that use the same aperture or antenna to transmit and receive energy, LIDAR sensors rely on a laser to transmit energy and separate optical sensor with a photon detector array (like those of EO/IR sensors) to collect the reflected light.[27]

LIDAR sensors have several advantages over EO/IR sensors:

- They can quickly obtain a precise target location including range and azimuth.
- Short laser pulses generate tight range resolutions that allow LIDAR sensors to separate targets from the fore- and background clutter.
- The same sensor can be used in both day and night conditions.

On the other hand, LIDAR suffers more severely from the effects of atmospheric attenuation, rain, fog, and haze because of the two-way transmission. Maximum laser power must be limited to make the lasers eye-safe when used near populations. And LIDAR sensors, too, will suffer from a high number of nuisance alarms, especially in degraded environments. Finally, because of their high angular and range resolutions, LIDAR sensors can generate millions of data points per second, requiring high levels of data processing. However, recent advances in both hardware and software have made such processing possible in near real time.

[27] There are bistatic radars that have separate transmitting and receiving antennas.

Example of LIDAR C-UAS Performance

Hammer et al. (2018) tested four scanning LIDAR systems working in conjunction against several sUAS including three quadcopters and one fixed-wing sUAS. All four vehicles had dimensions ranging from a quarter of a meter to one meter. All four LIDAR sensors, including two Velodyne HDL-64E and two Velodyne VLP-16 PUCK, operated from the same ground vehicle known as MODISSA. Both sensors rotated 360 degrees, generating 3D point clouds each containing 130K (HDL-64E) and 30K (VLP-16) point measurements. The horizontal/azimuthal resolutions of the HLD-64E and VLP-16 PUCK are 0.17 and 0.2 degrees when rotating at a rate of 10 Hz or 10 revolutions per second. Their vertical resolutions are 0.4 and 2.0 degrees, respectively. So, against a target at a range of 50 meters, the HLD-64E will have spatial sampling resolution of 0.15 by 0.35 meters, which is only slightly smaller than the target vehicles. Given their low resolution, these sensors can detect but cannot classify target sUASs even at relatively short ranges.

In this experiment, the Hammer et al. found that the LIDAR sensor package could reliably detect the target sUAS at ranges out to 30 meters, with some detections occurring out to 50 meters. False alarms increased significantly at ranges between 30 and 50 meters. It should be noted that both Velodyne systems are designed for applications like autonomous driving and not for long-range UAS detection. Higher-resolution LIDAR sensors that stare or scan more slowly may be able to extend detection ranges beyond what is presented here.

Threat Countermeasures Versus LIDAR

Physically smaller UASs naturally limit the number of LIDAR pulses placed on a target especially at longer ranges. And flying low and exploiting the terrain (e.g., trees, urban canyons) can block LIDAR energy from reaching the target, thereby reducing the probability of detection. Also, as was the case for radar sensors, there are camouflage techniques that absorb LIDAR energy in known spectral bands. The latter may be mitigated by higher-powered sensors operating across a wider spectral bandwidth. However, these may have limitations, especially when operating near civilian populations.

Summary of Findings from the Detectability Analysis

Effectively countering nefarious sUAS uses may require a range of sensor modalities, each with its own strengths and limitations. The analysis in this chapter led us to several observations about sUAS detectability and its challenges:

- To avoid being a nuisance, lawful operators may seek camouflage or reduced size to limit visual sensing or acoustic signature, trends that can provide advantages to nefarious operators as well.
- Although reduced RCS and command-and-control signal obfuscation may not be explicitly sought by lawful operators (and therefore market forces), they may appear as byproducts of other developments (e.g., airframe material or aircraft design and miniaturization affecting RCS, cell signal control unintentionally obfuscating operators).
- An effective systems-of-systems approach would be to search for and detect targets of interest using wide-area, lower-resolution sensors and then cue higher-resolution sensors to track and identify them. This requires a high degree of timely coordination among various sensors.
- sUASs are particularly problematic for C-UAS systems because, due to their size, they can generate many false alarms. This can play havoc on the human operators monitoring the sensors and systems, desensitizing them to the real threats and further limiting overall C-UAS effectiveness.
- The radar cross-section of rotary-wing sUASs can vary by several orders of magnitude, depending on the aspect angle of the sensor. This necessitates the use of multiple radars or a single radar with a wider bandwidth.
- Most military-grade target-tracking radars operate in the X-Band (8–12 GHz), where the antenna gain is high enough for good angular resolution, and, at the same time, the attenuation is weak enough to allow target tracking at ranges of hundreds of kilometers. However, against the sUAS threat, candidate radars do not have to detect and track at such long ranges, making higher-frequency bands (e.g., Ku, Ka) more attractive. One trade-off is that higher-frequency bands are more affected by poor weather conditions.
- Most low-cost, high-resolution EO/IR imagers with sufficient range have relatively small fields of view. As such, they will likely be most effective in detecting sUASs when cued by other sensors.
- Research has shown that some inexpensive thermal imagers are capable of detecting battery-powered sUASs.
- C-UAS systems that rely on ELINT are likely to be easily defeated by sUAS automation and autonomy.
- The combination of commercial market demand for reduced acoustic signatures and acoustic detection is limited to close ranges and is likely to minimize the utility of acoustic detection of sUASs in the future.
- LIDAR is unlikely to be useful for detecting sUASs in the near future. However, research has shown that some LIDAR systems are capable of detecting sUASs at ranges out to 30 meters.

CHAPTER SIX

Conclusions

Studying the performance characteristics of sUASs will help to inform DHS procurement decisions regarding C-UAS systems. This study has laid the groundwork for a living adversary-capability document regarding the range of potential nefarious sUAS systems, the technology they employ, their performance, their potential uses, and their detectability.

We identified four high-risk notional use cases:

1. unauthorized reconnaissance or surveillance
2. conveying illicit material
3. a kamikaze explosive (i.e., kinetic) attack
4. a CBR attack.

The four notional use cases developed for this study can be carried out using systems that are readily available on the commercial market today. At least 25 percent (351 out of 1,429) of the systems in the database could be used in one of the four use cases. At least 27 platforms on the market today can execute all four use cases; however, each system is specifically marketed and sold in the defense realm and therefore may not be available to retail consumers.

We also posited that expanding sUAS capabilities will only continue to challenge C-UAS efforts:

- Market estimates indicate that Chinese firm DJI accounts for 70 percent of the sUAS retail market. Trends in DJI system performance indicate rapid capability advancement. For example, the DJI Phantom's endurance has increased by about 1.5 minutes per year, and its maximum speed has increased 1.7 m/s over the past four years.
- Miniaturized radar, hyperspectral, and LIDAR payloads have been developed to work onboard sUASs. The Walden curve shows a tenfold increase in the processing capability of A/D converters, enabling sUASs to collect more electronic signals.[1]

[1] Again, the Walden curve describes the performance of systems, such as UAS sensors, that convert analog signals into digital signals.

- GNSS jamming and spoofing will continue to be credible applications for nefarious sUAS due to the limited power required to jam or spoof and the ability of an sUAS to get into unobstructed positions to maximize its impact.
- Much of the command-and-control market is focused on increasing the autonomy of flight software using artificial intelligence, predictive analytics, and computer vision to reduce the need for a highly trained operator.
- There are many how-to guides on the internet describing how to fit an sUAS with a cellular network transceiver for remote control. Users are likely dealing with latency on today's 4G network, but advances in this area will make it more difficult to identify sUASs via communication links.
- There is a robust open-source software and hardware ecosystem for supporting autonomous control, autopilot, and other sensors and control systems necessary to achieve remote operations.
- Increased autonomous capabilities and a desire for faster and simpler control are pushing manufacturers toward specialized controllers, often with separate controls for flight and payload. Autonomous technologies are reducing the complexity and need for operator training.
- Battery technology is notorious for overpromising, but we estimate that flow batteries could improve the range of the DJI Phantom 4, for example, by 20 percent, and solid-state batteries by 100 percent.

We additionally proposed a model by which a subset of sUASs could be identified based on the type of mission they are trying to perform. If DHS is interested in procuring C-UAS equipment to mitigate nefarious conveyance of material, for example, we can recommend 53 sUASs available today as a representative starting point for performance analysis. The specifics of any scenarios can necessitate the contraction or expansion of the sample, which is precisely why we designed the methods for generating samples of aircraft to be repeatable.

Furthermore, the data set was designed to be updated. AUVSI periodically updates its database, and it can be merged with additional data to enable the recomputation of a sample to support future analyses. Over time, the data set can be used to expand the trend analysis described in Chapter One.

Justifying acquisition decisions based on an understanding of the threat today can be misleading, particularly as technologies to continue to advance. To help address this problem, we also developed models of rotary-wing sUAS performance at a component level, enabling us to predict the performance of future sUASs. For example, we describe in Chapter Two how we predicted the performance of a hypothetical future DJI Phantom 4 with li-metal and flow batteries that could see a 20-percent increase in range.

We also began to enumerate the ways in which sensors are challenged in trying to detect, classify, identify, and track sUASs. This is why this report and its underlying data are intended to be foundational and updateable by S&T. Deeper understanding of technology capabilities and new insights about nefarious use cases and the implications for C-UAS can inform future analyses. This process was conceived of in three phases:

- Phase 1
  - Data collection: Build a consolidated data set of reported performance of commercial sUASs.
  - Historical data collection: Expand the consolidated data set with performance data over time to identify performance trends.
  - Analysis: Identify nefarious uses of sUASs.
  - Analysis: Describe reported performance characteristics (including trend analysis).
- Phase 2
  - Data collection: Build a data set of design characteristics (i.e., the components of sUASs and their performance).
  - Analysis: Incorporate existing and develop new models of performance for sUASs.
  - Analysis: Use models to predict performance and compare reported performance.
  - Analysis: Expand component-level analysis by creating hypothetical combinations of components to explore the capability of hypothetical systems.
- Phase 3
  - Data collection: Identify nefarious uses of sUASs and capture performance requirements (e.g., payload demands to carry out certain types of nefarious missions).
  - Analysis: Connect the phase 2 component-level analysis to the nefarious use requirements and determine the types and availability of systems and components that can execute nefarious acts.

This report presented the foundational phase 1 data development and analysis, some phase 2 component-level modeling and analysis, and some phase 3 integration of performance capabilities against nefarious uses. We recommend that future work update the phase 1 data, expand the model and component analysis in phase 2, look more holistically across the nefarious use cases, and assess in greater detail the performance demands of the use cases in phase 3.

References

Ackerman, Evan, "FlyJacket Lets You Control a Drone with Your Body," *IEEE Spectrum: Technology, Engineering, and Science News*, April 17, 2018. As of October 12, 2018:
https://spectrum.ieee.org/automaton/robotics/drones/epfl-flyjacket-exosuit-lets-you-control-a-drone-with-your-body

Adams, Eric, "Drop the Batteries—Diamonds and Lasers Could Power Your Drone," *Wired*, November 2018.

Amazon.com, "Sentry Portable ADS-B Receiver," product listing by ForeFlight, undated. As of October 12, 2018:
https://www.amazon.com/ForeFlight-Sentry-Portable-ADS-B-Receiver/dp/B07DVWH7BM

Andraši, Petar, Tomaslav Radišic, Mario Muštra, and Jurica Ivoševic, "Night-Time Detection of UAVs Using Thermal Infrared Camera," *Transportation Research Procedia*, Vol. 28, 2017, pp. 183–190.

Basile, Angelo, and Francesco Dalena, *Methanol: Science and Engineering*, Amsterdam, Netherlands: Elsevier, 2017.

Binnie, Jeremy, "Russians Reveal Details of UAV Swarm Attacks on Syrian Bases," *Jane's Defence Weekly*, January 12, 2018.

Borak, Masha, "World's Top Drone Seller DJI Made $2.7 Billion in 2017," TechNode, January 3, 2018. As of October 12, 2018:
https://technode.com/2018/01/03/worlds-top-drone-seller-dji-made-2-7-billion-2017

Brown, Daniel S., Michael A. Goodrich, Shin-Young Jung, and Sean C. Kerman, "Two Invariants of Human-Swarm Interaction," *Journal of Human-Robot Interaction*, Vol. 5, No. 1, 2016, pp. 1–31.

Brunning, Andy, "Periodic Graphics: Why Li-Ion Batteries Catch Fire," *Chemical and Engineering News*, Vol. 94, No. 45, November 14, 2016.

Buchmann, Isidor, "Bu-212: Future Batteries," webpage, Battery University, last updated May 31, 2018. As of October 23, 2018:
https://batteryuniversity.com/learn/article/experimental_rechargeable_batteries

Cambridge in Colour, "Lens Diffraction & Photography," undated. As of March 25, 2019:
https://www.cambridgeincolour.com/tutorials/diffraction-photography.htm

Chen, Chuan-Chung, *Attenuation of Electromagnetic Radiation by Haze, Fog, Clouds, and Rain*, Santa Monica, Calif.: RAND Corporation, R-1694-PR, 1975. As of March 25, 2019:
https://www.rand.org/pubs/reports/R1694.html

Cheng, Roger, "I Flew a Drone That Was 1,400 Miles Away and It Didn't Go So Well," *CNET*, March 3, 2016. As of October 12, 2018:
https://www.cnet.com/news/i-flew-a-drone-that-was-1400-miles-away-and-it-didnt-go-so-well

Choi, Jang Wook, and Doron Aurbach, "Promise and Reality of Post-Lithium-Ion Batteries with High Energy Densities," *Nature Reviews Materials*, Vol. 1, Article 16013, April 2016.

Clancy, Joseph, Director, U.S. Secret Service, written testimony at the hearing "Flying Under the Radar: Securing Washington D.C. Airspace," Committee on Oversight and Government Reform, U.S. House of Representatives, 113th Congress, 2nd Session, April 29, 2014. As of October 31, 2018:
https://www.dhs.gov/news/2014/04/29/written-testimony-usss-director-house-committee-oversight-and-government-reform

Corfield, Gareth, "GitHub Shrugs Off Drone Maker DJI's Crypto Key DMCA Takedown Effort," *The Register*, January 25, 2018. As of October 12, 2018:
https://www.theregister.co.uk/2018/01/25/dji_github_public_repo_crypto_key_foolishness

Darrah, M. R., A. K. Raj, and S. V. Drakunov, "A Fractal Control Architecture for Multiple UAVs Missions," paper presented at the AUVSI Unmanned Systems North America Conference, 2008.

Dey, Vishal, Vikramkumar Pudi, Anupam Chattopadhyay, and Yuval Elovici, "Security Vulnerabilities of Unmanned Aerial Vehicles and Countermeasures: An Experimental Study," *2018 31th International Conference on VLSI Design and 2018 17th International Conference on Embedded Systems*, IEEE, March 2018.

Drone II, "TOP20 Drone Company Rankings, Q3 2016," September 2016. As of October 12, 2018:
https://www.droneii.com/project/top20-drone-company-ranking-q3-2016

"Drone Technology Has Made Huge Strides," *The Economist*, June 10, 2017.

Echodyne, "EchoFlight Radar," web page, undated. As of March 25, 2019:
https://www.echodyne.com/products/echoflight

Empower RF Systems, Inc., "Solid State Broadband High Power Amplifier," product sheet, April 18, 2016. As of October 12, 2018:
https://www.empowerrf.com/datasheet/Empower_RF_Amplifier_1164.pdf

Esc Aerospace, "HAES 400 Aerial Target," fact sheet, 2012. As of October 15, 2018:
http://www.evolvsys.cz/files/flyers/UAS_HAES_400.pdf

Etherium, homepage, 2018. As of September 4, 2018:
https://www.ethereum.org

FAA—*See* Federal Aviation Administration.

Farlik, Jan, Miroslav Kratky, Josef Casar, and Vadim Stary, "Radar Cross Section and Detection of Small Unmanned Aerial Vehicles," paper presented at Mechatronika: 17th International Conference on Mechatronics, Prague, Czech Republic, December 2016.

Federal Aviation Administration, "Fact Sheet—Small Unmanned Aircraft Regulations (Part 107)," July 23, 2018. As of March 25, 2019:
https://www.faa.gov/news/fact_sheets/news_story.cfm?newsId=22615

Federal Protective Service, *Federal Protective Service Annual Report, Fiscal Year 2015*, 2016. As of October 31, 2018:
https://www.dhs.gov/sites/default/files/publications/Federal%20Protective%20Service%20Annual%20Report%20508%20Compliant%20FY2015.pdf

FOAM, homepage, undated. As of September 4th, 2018:
https://foam.space

———, *FOAM Technical Whitepaper*, draft 0.4, 2018a.

———, *FOAM Whitepaper*, May 1, 2018b.

FPS—*See* Federal Protective Service.

Fu, Yangqing, Morgann Kinniry, and Laura N. Kloepper, "The Chirocopter: A UAV for Recording Sound and Video of Bats at Altitude," *Methods in Ecology and Evolution*, Vol. 9, No. 6, June 2018, pp. 1531–1535.

Giangreco, Leigh, "DARPA Completes Second Phase of Swarming Demo," *FlightGlobal*, January 12, 2018. As of October 12, 2018:
https://www.flightglobal.com/news/articles/darpa-completes-second-phase-of-swarming-demo-444863

Gibbons-Neff, Thomas, "Watch the Pentagon's New Hive-Mind-Controlled Drone Swarm in Action," *Washington Post*, January 10, 2017.

Gonzalez, Robbie, "How a Flock of Drones Developed Collective Intelligence," *Wired*, July 18, 2018. As of March 22, 2019:
https://www.wired.com/story/how-a-flock-of-drones-developed-collective-intelligence

Goode, Lauren, "Batteries Still Suck, but Researchers Are Working on It," *Wired*, May 22, 2018. As of October 12, 2018:
https://www.wired.com/story/building-a-better-battery

Goodrich, Michael A., Sean Kerman, Brian Pendleton, and P. B. Sujit, "What Types of Interactions Do Bio-Inspired Robot Swarms and Flocks Afford a Human?" *Proceedings of Robotics: Science and Systems*, Cambridge, Mass.: MIT Press, 2013, pp. 105–112.

Gorey, Colm, "MIT May Have Just Solved How to Mass-Produce Graphene," *Silicon Republic*, April 19, 2018. As of March 22, 2019:
https://www.siliconrepublic.com/machines/mass-produce-graphene-solved

Griebel, Joseph Charles Pius, "Scaling Limitations of Micro Engines," *Undergraduate Research Journal at University of Colorado at Colorado Springs*, Vol. 2, No. 3, 2010.

Hammer, Marcus, Marcus Hebel, Bjorn Borgmann, Martin Laurenzis, and Michael Arens, "Potential of Lidar Sensors for the Detection of UAVs," *Proceedings of the SPIE*, Vol. 10636, May 10, 2018.

Harju, Ari, Topi Siro, Filippo Federici Canova, Samuli Hakala, and Teemu Rantalaiho, "Computation Physics on Graphics Processing Units," *Proceedings of the 11th International Conference on Applied Parallel and Scientific Computing*, Berlin: Springer-Verlag, 2012, pp. 3–26.

Hill, Kashmir, "Jamming GPS Signals Is Illegal, Dangerous, Cheap, and Easy," *Gizmodo*, July 24, 2017. As of October 12, 2018:
https://gizmodo.com/jamming-gps-signals-is-illegal-dangerous-cheap-and-e-1796778955

Hinnrichs, Michele, Bradford Hinnrichs, and Earl McCutchen, "Infrared Hyperspectral Imaging Miniaturized for UAV Applications," *Proceedings of the SPIE*, Vol. 10177, May 9, 2017.

Homakov, Egor, "Stop. Calling. Bitcoin. Decentralized," *Medium*, December 2, 2017. As of October 12, 2018:
https://medium.com/@homakov/stop-calling-bitcoin-decentralized-cb703d69dc27

Hosenball, Mark, "Security High as Phoenix Prepares for Super Bowl," Reuters, January 23, 2015. As of October 31, 2018:
https://www.reuters.com/article/us-nfl-superbowl-security/security-high-as-phoenix-prepares-for-super-bowl-idUSKBN0KW2GB20150123

Huizing, A.G. et al., "Compact Scalable Multifunction RF Payload for UAVs with Frequency Modulated Continuous Wave (FMCW) Radar and Electronic Support Measures (ESM) Functionality," *Proceedings of the Radar Conference*, IEEE, November 2009.

Impossible Aerospace, homepage, undated. As of October 12, 2018:
https://impossible.aero

International Telecommunications Union, *Specific Attenuation Model for Rain for Use in Prediction Methods*, Recommendation ITU-R P.838, March 8, 2005. As of March 22, 2019:
https://www.itu.int/rec/R-REC-P.838-3-200503-I/en

Irving, Michael, "DARPA Program to Allow for Mid-Flight Multitasking Drone Missions," *New Atlas*, June 7, 2017. As of October 12, 2018:
https://newatlas.com/darpa-bae-systems-uas-multifunction/49931/

Jammer-Store, "GJ6 Portable All Civil Bands GPS Jammer, Anti Tracking Device," product listing, undated. As of October 12, 2018:
https://www.jammer-store.com/gj6-all-civil-gps-signal-jammer-blocker.html

Jülich Forschungszentrum, "It's All in the Mix: Jülich Researchers Are Developing Fast-Charging Solid-State Batteries," press release, August 20, 2018. As of October 12, 2018:
http://www.fz-juelich.de/SharedDocs/Pressemitteilungen/UK/EN/2018/2018-08-20-its-all-in-the-mix--fast-charging-solid-state-batteries.html

Kapoor, Rohan, Subramanian Ramasamy, Alessandro Gardi, and Roberto Sabatini, "A Bio-Inspired Acoustic Sensor System for UAS Navigation and Tracking," *Proceedings of the 2017 IEEE/AIAA 36th Digital Avionics Systems Conference (DASC)*, November 9, 2017.

Kato, Yuki, Satoshi Hori, Toshiya Saito, Kota Suzuki, Masaaki Hirayama, Akio Mitsui, Masao Yonemura, Hideki Iba, and Ryoji Kanno, "High-Power All-Solid-State Batteries Using Sulfide Superionic Conductors," *Nature*, March 21, 2016.

Kloet, N., S. Watkins, and R. Clothier, "Acoustic Signature Measurement of Small Multi-Rotor Unmanned Aircraft Systems," *International Journal of Micro Air Vehicles*, Vol. 9, No. 1, February 2017, pp. 3–14.

Kolling, Andreas, Katia Sycara, Steven Nunnally, and Michael Lewis, "Human-Swarm Interaction: An Experimental Study of Two Types of Interaction with Foraging Swarms," *Journal of Human-Robot Interaction*, Vol. 2, No. 2, June 2013 pp. 103–129.

Kolling, Andreas, Phillip Walker, Nilanjan Chackraborty, Katia Sycara, and Michael Lewis, "Human Interaction with Robot Swarms: A Survey," *IEEE Transactions on Human-Machine Systems*, Vol. 46, No. 1, February 2016, pp. 9–26.

Koretsky, G. M., J. F. Nicoll, and M. S. Taylor, *Tutorial on Electro-Optical/Infrared (EO/IR) Theory and Systems*, Alexandria, Va.: Institute for Defense Analyses, January 2013.

Li, Charlie, "Make Your Personal Drone Fly Even Farther with a 4G Network Connection," blog post, *Wiredcraft*, May 3, 2016. As of October 12, 2018:
https://wiredcraft.com/blog/drone-copter-uav-4g-network

Lichtman, March, Raghunandan Rao, Vuk Marojevic, Jeffrey Reed, and Roger Piqueras Jover, "5G Jamming, Spoofing, and Sniffing: Threat Assessment and Mitigation," revised April 8, 2018. As of October 12, 2018:
https://arxiv.org/abs/1803.03845

Lindeburg, Michael R., *Mechanical Engineering Reference Manual for the PE Exam*, 9th ed., Belmont, Calif.: Professional Publications, 1995.

LoRa Alliance, "What Is the LoRaWAN™ Specification?" webpage, undated. As of September 4, 2018:
https://lora-alliance.org/about-lorawan

Lowe, Derek, "Graphene: You Don't Get What You Pay For," blog post, Science Translational Medicine, October 11, 2018. As of March 25, 2019:
http://blogs.sciencemag.org/pipeline/archives/2018/10/11/graphene-you-dont-get-what-you-pay-for

Masterson, Andrew, "Autonomous and Cooperating: The Dawn of the Drone Swarm," *Cosmos Magazine*, July 19, 2018. As of March 22, 2019:
https://cosmosmagazine.com/technology/
autonomous-and-cooperating-the-dawn-of-the-drone-swarm

MathWorks, "Modeling Target Radar Cross Section," webpage, undated. As of March 25, 2019:
https://www.mathworks.com/help/phased/examples/modeling-target-radar-cross-section.html

MatWeb, "Property Search," webpage, undated. As of March 22, 2019:
http://matweb.com/search/PropertySearch.aspx

McLees, Andrew, Immigration and Customs Enforcement, testimony at the hearing "Mass Gathering Security: A Look at the Coordinated Approach to Super Bowl XLVIII in New Jersey and Other Large-Scale Events," Committee on Homeland Security, U.S. House of Representatives, 113th Congress, 2nd Session, June 23, 2014. As of October 31, 2018:
https://www.gpo.gov/fdsys/pkg/CHRG-113hhrg90883/html/CHRG-113hhrg90883.htm

Mekonnen, Yemeserach, Aditya Sundararajan, and Arif I. Sarwat, "A Review of Cathode and Anode Materials for Lithium-Ion Batteries," paper presented at SoutheastCon 2016, Norfolk, Va., March 30, 2016.

Mesbahi, Nabil, and Hamza Dahmouni, "Delay and Jitter Analysis in LTE Networks," *Proceedings of the 2016 International Conference on Wireless Networks and Mobile Communications (WINCOM)*, IEEE, December 2016.

Montroll, Mark, *Capability-Based Acquisition in the Missile Defense Agency*, Ft. McNair, Washington, D.C.: Industrial College of the Armed Forces, National Defense University, 2003. As of October 12, 2018:
http://www.dtic.mil/dtic/tr/fulltext/u2/a422428.pdf

Morgasinski, A., P. Dixon, B. Hertzberg, A. Kvit, J. Ayala, and G. Yushin, "High-Performance Lithium-Ion Anodes Using a Hierarchical Bottom-Up Approach," *Nature Materials*, Vol. 9, No. 4, April 2010, pp. 353–358.

Moses, Allistair, Matthew J. Rutherford, Michail Kontitsis, and Kimon P. Valavanis, "Miniature UAV Radar System," briefing, undated.

Murmann, B., "ADC Performance Survey, 1997–2019," spreadsheet, Stanford, Calif.: Stanford University, 2018. As of October 12, 2018:
https://web.stanford.edu/~murmann/adcsurvey.html

NanoFlowcell, homepage, undated. As of October 12, 2018:
https://www.nanoflowcell.com

NASA—*See* National Aeronautics and Space Administration.

National Academies of Sciences, Engineering, and Medicine, *Counter-Unmanned Aircraft System (CUAS) Capability for Battalion-and-Below Operations: Abbreviated Version of a Restricted Report*, Washington, D.C.: National Academies Press, 2018.

National Aeronautics and Space Administration, "Unmanned Aircraft System (UAS) Traffic Management (UTM)," webpage, last updated September 25, 2018. As of October 12, 2018:
https://utm.arc.nasa.gov/index.shtml

Nave, Rod, "Diffraction-Limited Imaging," *HyperPhysics Concepts*, Georgia State University, 2016a. As of March 25, 2019:
http://hyperphysics.phy-astr.gsu.edu/hbase/phyopt/diflim.html

———, "The Rayleigh Criterion," *HyperPhysics Concepts*, Georgia State University, 2016b. As of March 25, 2019:
http://hyperphysics.phy-astr.gsu.edu/hbase/phyopt/Raylei.html

Nicholas, Robert, Ronald Driggers, David Shelton, and Orges Furxhi, "Infrared Search and Track Performance Estimates for Detection of Commercial Unmanned Aerial Vehicles," *Proceedings of the SPIE*, Vol. 10625, April 2018.

Pixhawk, homepage, undated. As of October 12, 2018:
http://pixhawk.org

Porat, Talya, Tal Oron-Gilad, Michal Rottem-Hovev, and Jacob Silbiger, "Supervising and Controlling Unmanned Systems: A Multi-Phase Study with Subject Matter Experts," *Frontiers in Psychology*, Vol. 7, May 24, 2016.

PX4, *PX4 Autopilot User Guide*, October 11, 2018. As of October 12, 2018:
http://docs.px4.io/en

Reese, Shawn, "National Security Special Events: Fact Sheet," Washington, D.C.: Congressional Research Service, January 25, 2017.

Ritchie, Matthew, Francesco Fioranelli, Hugh Griffiths, and Borge Torvik, "Micro-Drone Analysis," *Proceedings of the 2015 IEEE Radar Conference*, June 2016.

Semtech, "What Is LoRa®," webpage, undated. As of September 4, 2018:
https://www.semtech.com/lora/what-is-lora

SESAR Joint Undertaking, "U-Space," webpage, undated. As of October 12, 2018:
https://www.sesarju.eu/U-Space

Sher, E., and I. Sher, "Theoretical Limits of Scaling-Down Internal Combustion Engines," *Chemical Engineering Science*, Vol. 66, No. 3, February 2011, pp. 260–267.

Shotspotter, "Shotspotter Fact Sheet," undated. As of March 25, 2019:
http://www.shotspotter.com/system/content-uploads/ShotSpotter_Fact_Sheet_-_final_draft_12.13.pdf

Sims, Brendan, and Simon Crase, *Review of Battery Technologies for Military Land Vehicles*, Australian Department of Defence, January 2017.

Sion Power Corporation, "Sion Power's Lithium-Sulfur Batteries Power World Record Flight," press release, July 29, 2010. As of October 22, 2018:
https://www.businesswire.com/news/home/20100729006618/en/Sion-Power%E2%80%99s-Lithium-Sulfur-Batteries-Power-World-Record

SLAC National Accelerator Laboratory, "X-Rays Uncover a Hidden Property That Leads to a Failure in a Lithium-Ion Battery Material," press release, September 17, 2018. As of October 12, 2018:
https://www6.slac.stanford.edu/news/2018-09-17-x-rays-uncover-hidden-property-leads-failure-lithium-ion-battery-material.aspx

Smith, Cassie, "3D Printing Trends to Watch in 2018," *CAD Crowd*, October 10, 2018. As of March 22, 2019:
https://www.cadcrowd.com/blog/3d-printing-trends-to-watch-in-2018

Sneiderman, Phil, "Johns Hopkins Scientists Show How Easy It Is to Hack a Drone and Crash It," press release, Johns Hopkins University, June 8, 2016.

Squarehead Technology, "Discovair Rugged Acoustic Drone Detection," undated. As of December 2018:
http://www.sqhead.com/drone-detection

Sullivan, John P., Robert J. Bunker, and David A. Kuhn, "Mexican Cartel Tactical Note #38: Armed Drone Targets the Baja California Public Safety Secretary's Residence in Tecate, Mexico," *Small Wars Journal*, August 6, 2018. As of October 31, 2018:
http://smallwarsjournal.com/jrnl/art/mexican-cartel-tactical-note-38-armed-drone-targets-baja-california-public-safety

Suo, Liumin M., Oleg Borodin, Tao Gao, Marco Olguin, Janet Ho, Xuilin Fan, Chao Luo, Chunsheng Wang, and Kang Xu, "'Water-in-Salt' Electrolyte Enables High-Voltage Aqueous Lithium-Ion Chemistries," *Science*, Vol. 350, No. 6263, November 20, 2015, pp. 938–943.

Temple, James, "This Battery Advance Could Make Electric Vehicles Far Cheaper," *MIT Technology Review*, April 11, 2018. As of October 12, 2018:
https://www.technologyreview.com/s/610792/
this-battery-advance-could-make-electric-vehicles-far-cheaper

The Things Network, "LoRaWAN Overview," webpage, undated. As of September 4, 2018:
https://www.thethingsnetwork.org/docs/lorawan

Trujano, Fernando, Benjamin Chan, Greg Beams, and Reece Rivera, *Security Analysis of DJI Phantom 3 Standard*, Cambridge, Mass.: Massachusetts Institute of Technology, May 11, 2016. As of October 12, 2018:
https://courses.csail.mit.edu/6.857/2016/files/9.pdf

UASweekly, "Ukrainian Company Unveils New Drone with Grenade Launcher," August 20, 2018. As of October 31, 2018:
http://uasweekly.com/2018/08/20/ukrainian-company-unveils-new-drone-with-grenade-launcher

UgCS, "Drone Show Software," webpage, undated. As of March 25, 2019:
https://www.ugcs.com/page/droneshowsoftware

Unmanned Systems Technology, "Horizon Energy System News," webpage, undated. As of October 12, 2018:
https://www.unmannedsystemstechnology.com/tag/horizon-energy-systems/

Urbonavicius, Marius, Sarunas Varnagiris, Dalius Girdzevicius, and Darius Milcius, "Hydrogen Generation Based on Aluminum-Water Reaction for Fuel Cell Applications," *Energy Procedia*, Vol. 128, September 2017, pp. 114–120.

U.S. Department of State, *Trafficking in Persons 2018 Report: Country Narratives*, Washington, D.C., 2018, pp. 84–86. As of October 31, 2018:
https://www.state.gov/j/tip/rls/tiprpt/2018

Vásárhelyi, Gábor, Csaba Virágh, Gergő Somorjai, Tamás Nepusz, Agoston E. Eiben, and Tamás Vicsek, "Optimized Flocking of Autonomous Drones in Confined Environments," *Science Robotics*, Vol. 3, No. 20, July 18, 2018.

Walker, Phillip, Saman Amirpour Amraii, Nilanjan Chakraborty, Michael Lewis, and Katia Sycara, "Human Control of Robot Swarms with Dynamic Leaders," paper presented at the 2014 IEEE/RSJ International Conference on Intelligent Robots and Systems, Chicago, Ill., September 14–18, 2014.

Walters, Sander, "How Can Drones Be Hacked? The Updated List of Vulnerable Drones and Attack Tools," *Medium*, October 29, 2016. As of October 12, 2018:
https://medium.com/@swalters/how-can-drones-be-hacked-the-updated-list-of-vulnerable-drones-attack-tools-dd2e006d6809

Warwick, Graham, "Liquid Battery Promises Safe Energy-Dense Power for Electric Aircraft," *Aviation Week and Space Technology*, August 14, 2018. As of October 12, 2018:
http://aviationweek.com/future-aerospace/liquid-battery-promises-safe-energy-dense-power-electric-aircraft

Wikipedia, "Link 16," last updated September 26, 2018b. As of October 12, 2018:
https://en.wikipedia.org/wiki/Link_16

Wolf, Christian, "Phased Array Antenna," webpage, undated. As of October 12, 2018:
http://www.radartutorial.eu/06.antennas/Phased%20Array%20Antenna.en.html

Xcelerit, "Benchmarks: Intel Scalable CPU vs. Nvidia V100 GPU," July 9, 2018. As of March 25, 2019:
https://www.xcelerit.com/computing-benchmarks/insights/benchmarks-intel-xeon-scalable-processor-vs-nvidia-v100-gpu

Xia, C., C. Y. Kwok, and L. F. Nazar, "A High-Energy-Density Lithium-Oxygen Battery Based on a Reversible Four-Electron Conversion to Lithium Oxide," *Science*, Vol. 361, No. 6404, August 24, 2018, pp. 777–781.

Yoon, Ginsu, "Magnificent Seven," *Ginsudo*, January 5, 2016. As of April 4, 2019:
https://blog.ginsudo.com/2016/01

Zeng, Yong, Jiangbin Lyu, and Rui Zhang, "Cellular-Connected UAV: Potentials, Challenges and Promising Technologies," submitted April 6, 2018. As of October 12, 2018:
https://arxiv.org/abs/1804.02217